University of Plymouth Library

Subject to status this item may be renewed
via your Voyager account

http://voyager.plymouth.ac.uk

Exeter tel: (01392) 475049
Exmouth tel: (01395) 255331
Plymouth tel: (01752) 232323

Safety Factors and Reliability: Friends or Foes?

Safety Factors and Reliability: Friends or Foes?

by

ISAAC ELISHAKOFF

J.M. Rubin Professor of Structural Reliability, Safety and Security,
Department of Mechanical Engineering,
Florida Atlantic University,
Boca Raton, Florida, U.S.A.

KLUWER ACADEMIC PUBLISHERS

DORDRECHT / BOSTON / LONDON

A C.I.P. Catalogue record for this book is available from the Library of Congress.

ISBN 1-4020-1779-0 (HB)
ISBN 1-4020-2131-3 (e-book)

Published by Kluwer Academic Publishers,
P.O. Box 17, 3300 AA Dordrecht, The Netherlands.

Sold and distributed in North, Central and South America
by Kluwer Academic Publishers,
101 Philip Drive, Norwell, MA 02061, U.S.A.

In all other countries, sold and distributed
by Kluwer Academic Publishers,
P.O. Box 322, 3300 AH Dordrecht, The Netherlands.

Printed on acid-free paper

Printed in the Netherlands.

Dedicated to the blessed memory
of my beloved brother Moshe

Contents

x

Chapter 1
Prologue

"The safety of the building constructions is a matter of calculating probabilities."

M. Mayer (1926)

"The values of safety factors, as well as closely associated values of design loads and design resistances, were improved and modified mainly empirically, by the way of generalization of multi-year experience and exploitation of the structures. Yet, as is seen of the essence of the problem in principle, there are also theoretical approaches possible with wide application of the apparatus of theory of probability and mathematical statistics."

V.V. Bolotin (1965)

"Probability theory provides a more accurate engineering representation of reality. Many leading civil engineers in many countries have written of the statistical nature of loads and of material properties."

C.A. Cornell (1969)

"The times of straightforward structural design, when the structural engineer could afford to be fully ignorant of probabilistic approaches to analysis, are definitely over."

A.M. Lovelace (1972)

"As a person who was brought up on factors of safety and used them all his professional life, their simplicity appeals to me. However, if we are to make any progress the bundle has to be unbundled, and each of the constituents correctly modeled..."

A.D.S. Carter (1997)

It is a conventional wisdom to maintain that the scientists and engineers, who earn a living by being engaged in applied mechanics, are divided into two groups: those in the first, traditional group deal with deterministic mechanics, whereas the representatives of the second, more recent, group devote themselves to non-deterministic mechanics. The traditionalists neglect uncertainties in the loading conditions, in the mechanical properties of structures, in boundary conditions, in geometric characteristics and in other parameters entering into the description of the problem at hand.

The second group embraces itself with various analyses of uncertainty. Within this group, there are those who are active in probabilistic mechanics. They maintain that uncertainty is identifiable with randomness, and hence methods of classical or modern probability theories should be applied. Other investigators developed fuzzy sets based theories, formulating their analysis on the principle developed by Lofti Zadeh:

1

"As the complexity of a system increases, our ability to make precise and yet significant statements about its behavior diminishes until a threshold is reached beyond which precision and significance (or relevance) become almost mutually exclusive characteristics."

The theorists utilizing the fuzzy sets approach base their approach on the notion of the membership function, in contrast to the concept of the probability density function utilized by probabilists. Whereas the two approaches seem to be radically different, they have one thing in common: They both use an uncertainty measure. Engineers developed also an uncertainty analysis that does not demand a measure. The latter method is known by its various appellations: method of accumulation of disturbances, unknown-but-bounded uncertainty, convex modeling of uncertainty, anti-optimization and more recently information-gap methodology.

The set of scientists belonging to the groups dealing with the deterministic mechanics or with the non-deterministic mechanics are not mutually exclusive. A very limited number of scholars simultaneously deal with both approaches. Even smaller numbers of researchers are engaged in seemingly competitive uncertainty analyses techniques. Worldwide, these investigators apparently can be counted on the fingers of one hand. The latter researchers appear to exercise tolerance by holding several contradicting opinions since they find similarities or analogies between above contradictory approaches, indeed; according to F.S. Fitzgerald,

"the test of a first-rate intelligence is the ability to hold to opposite ideas in the mind at the same time, and still retain the ability to function."

The main goal of this prologue is not a classification of researches in applied mechanics, but rather, to demonstrate that the above characterization, in essence, is imprecise if not altogether wrong. The main thesis that we would like to propagate is this: In actuality there is not such thing as deterministic mechanics.

Such a claim may appear, in the superficial reading, to be highly controversial. But once we review the main premises of what is known as a deterministic mechanics, we will acknowledge that the above statement is quite transparent. Indeed, although it is assumed within the deterministic paradigm, that uncertainty is absent, at the latest stage of the deterministic analysis, after stresses, deformations and displacements are found by quite sophisticated analytical and/or numerical techniques, somehow, and nearly miraculously, the neglected uncertainties are taken into account. These uncertainties are enveloped by the concept of SAFETY FACTOR. Thus, the uncertainty is introduced *via the back door*.

According to Vanmarke (1979),

"the format of many existing codes and design specifications impedes rational risk assessment or communication among designers and owners about risk. For example, the conventional safety factor of safety format considers a design acceptable if the computed factor of safety exceeds prescribed allowable value. Such a criterion characterizes structures as either safe or unsafe. It leads many engineers to embracing the concept that all alternative designs which satisfy the criteria are absolutely safe. Consequently, a little thought is given to the ever-present probability of failure, to the factors which influence it, and to opportunities for providing added protection to reduce the risk of failure for better or worse, rigid design provisions such as the factor-of-safety format take much of the responsibility for decision making out of the hands of the engineer. In this sense, codes often serve as cookbooks or as crutches."

According to Ditlevsen (1981),

"..the inherent property of engineering quantities remains outside the traditional system of engineering. The system is customarily represented as being deterministic; to each quantity, it is presumed, a unique value can be assigned (e.g., the yield strength of structural steel is $260 \, N/mm^2$, 100 vehicles per hour, make a left turn at the intersection, etc.). In turn structures, pipe networks, earthworks, etc., are conventionally modeled to behave in a unique mechanistic way for a given set of the model's parameter values (dimensions, etc.) and for a given set of input quantities (loads, applied pressures, etc.). The conclusion is that their responses are predictable with certainty."

The inquisitive reader would ask: "What are the reasons embracing the deterministic mechanics and excluding uncertainty?" Ditlevsen (1981) provides the following insight to this possible inquiry: "The reasons for this exclusion have been clear enough: the deterministic representation is easier to learn and use..."

Haugen (1980) writes:

"To design is to formulate a plan for the fulfillment of a human need "(Shigley, 1977).

Krick (1967) considers design a "decision-making process." Initially, the need for design may be well defined; however, the problem often be somewhat nebulous. There is now a choice of philosophies available for carrying out mechanical design: (1) design based on theory of probability and (2) design based on deterministic assumptions."

This book deals with the following questions: Can uncertainty be introduced *via the front door*? Can the safety factor and the measures characterizing the non-deterministic considerations coexist peacefully? We hope that some partial answers will be provided to these non-trivial questions.

One fact needs to be emphasized. It also spectacularly distinguishes between deterministic and non-deterministic philosophies. While the deterministic method claims that once the failure criteria, in conjunction with the safety factor are satisfied, the structure is absolutely immune to failure, the non-deterministic probabilistic approach retains the possibility of failure, unless the probability density is zero outside a finite interval. Thus, the non-deterministic approach appears to be more "honest", putting all consequences on the table, instead of the deterministic approach which discards the failure by relegating it, to a place "under the rug."

The Rambam–Rabbi Moshe Ben Maimon (Moses Maimonides) (born in Cordoba, Spain on March 30, 1135, died in Fostat, Egypt on December 13, 1204; buried in Israel)

Haugen stresses (1980):

"Since safety factors are not performance-related measures, there is no way by which an engineer can know whether his designs are near optimum or overly conservative. In many instances, this may not be important, but in others, it can be critical (Roark, 1965)."

Still we can safely claim that the person who suggested the notion of the safety factor was a genius. This factor allowed and continues to enable constructing safe or nearly safe structures that work. "If it works, use it", one would say. The question is: Could such a methodology be improved? Can we do it better, even though the American proverb advises, "if it ain't broke, don't fix it."?

This monograph maintains that the concept of reliability –probability that the structure will perform its intended mission– can be used to enhance the deterministic approach based on safety factors. Can we augment the safety factor based design to remove the mystery from it? The mystery is, of course, in choosing some number, out of the sky, be it 1.2, 1.7 or 2.3 or other that must guarantee the structure's safety? Some justification in choosing the number may also provide a key for the choice of rigorous method(s).

The origins of the concept of safety factor go to times immemorial. You possibly guessed it! The main idea apparently is expressed in the *Torah* (also known as the "Bible"), Deuteronomy 22-8:

> "If you build a new house, you should make a fence for your roof, so that you will not place blood in your house if a fallen one falls from it."

This fence from the dangerous edge removes occupants of the house from the possible accident. The Stone edition of the *Torah* (the "Bible") mentions (Sherman, 1993):

> "According to Rambam, also known as Maimonides, a Jewish sage and philosopher of the Twelve's century, this commandment applies to any dangerous situation, such as a swimming pool or a tall stairway (see Rambam, *Hil. Rotz.* 11: 1-5)"

Likewise, engineers put a numerical "fence" below the level of the yield stress. The latter property –that of increase of deformation without increase in the stress level– is easily identifiable with a dangerous situation. Yield stress "warns" us, against eminent danger, and "asks" to install the fence. At this juncture it is instructive to remember the quote by Albert Einstein: "The most incomprehensible thing about the world is that it is comprehensible" (see Gorson-Milgrim, 1962). This is how, quite possibly, we could speculate about the origins of the safety factor. Materials with the yield stress appear to be perfectly designed to introduce the fence around it. Yield stress, therefore, is both a danger and a blessing: It is easy to recognize that the yield condition is a danger, but it is less trivial to recognize it as a blessing.

One comment is needed here: Moses Maimonides stresses the need to distance oneself from a dangerous situation. In mechanics of solids we may not have yield stress level alone as a indicator of the dangerous situation, some materials exhibit an ultimate stress instead. Therefore, it makes sense to talk about *resistance* or *capacity* that should not be exceeded by the structure failure criterion.

Hart (1982, p.118) first poses a question about defining a failure and then replies to it:

> "What is *failure*? Failure is what the structural engineer defines it to be and nothing else. For example, of the stress induced by an earthquake exceeds the yield stress of the material, it could be called failure. Alternatively, if the stress exceeds the ultimate stress of the material, it could be called failure. Failure can also be related to structure serviceability…. Therefore, it is fundamentally important to realize that the structural engineer defines failure and that the examples are virtually unlimited."

According to Gnedenko et al (1969),

> "A *failure* is the partial or total loss or modification of those properties of the units in such a way that their functioning is seriously impeded or completely stopped. In certain cases, the concept of failure is sharply defined. A typical example of a component having a well-defined failure is an electric light bulb. The operation of a light bulb has, as a rule, two states: either it gives normal

illumination or it gives no illumination at all. However, in connection with electronic units, the concept of failure is extremely relative since it depends in a significant way on the particular conditions under which the unit may be used."

Charles Augustin Coulomb (born on June 14, 1736 in Angouleme, France; died on August 23, 1806 in Paris, France)

Petroski (1996) comments about the concept of failure:

"An idea that unifies all of engineering is the concept of failure. From the simplest paper clips to the finest pencil leads to the smoothest operating zippers, inventions are successful only to the extent that their creators properly anticipate how a device can fail to perform as intended. Virtually every calculation that an engineer can perform in the development of computers and airplanes, or telescopes and fax machines, is a failure calculation…What distinguishes the engineer from the technician is largely the ability to formulate and carry out the detailed calculation of forces and deflections, concentrations and flows, voltages and currents, that are required to test proposed design on paper with regard to failure criterion."

According to Casciati (1991) failure includes

" – loss of static equilibrium of the structure, or a part of the structure, considered as a rigid body,

- localized rupture of critical sections of the structure caused by exceeding the ultimate strength (possibly reduced by repeated loading), on the ultimate deformation of the material,
- transformation of the structure into a mechanism,
- general or local instability,
- progressive collapse,
- deformation which affects the efficient use or appearance of structural or non-structural elements,
- excessive vibration, which may cause discomfort and/or alarm,
- local damage (including cracking), which affects the durability of the efficiency of the structure."

Gertsbakh and Kordonsky (1969) classify the reasons for failure as follows:

"Construction Defects. Failures of this group arise as a consequence of an imperfection in its construction. A typical example is non-consideration of "peak" loads.

A load acting on a system and its elements usually has random variations. In the construction, one tries to keep in mind the possibility of occurrence of "peak" loads, that is, loads considerably exceeding the loads due to normal use. If an analysis and calculation of the loads are made with insufficient care, then the action of "peak" loads will lead to failure. From this point of view of analysis and calculation of reliability, it is important to have the defects in the construction show up to the same extent on all copies of the system or element under consideration.

Technological defects. Failures of this class occur as a consequence of violation of the technological manufacturing procedure chosen for the system or unit. The quality of the individual units and connections and of a unit as a whole has unavoidable random variations. Quality variations kept within sufficiently restricted limits do not show up appreciably in the reliability of the system. With sharp fluctuations in the quality, the reliability of certain items will prove considerably less than the reliability of others. Therefore, technological defects decrease the reliability of some of the items in the total set of manufactured systems or units.

Claude Louis Marie Henri Navier (born on February 10, 1785 in Dijon, France; died on August 21, 1836 in Paris, France)

<u>Defects due to improper use.</u> For every system, restrictions are made on the conditions of its use (restrictions on the temperature, on frequency of vibration, etc.) and rules are given for maintenance of the system and its parts, and so forth. Violation of the rules of use leads to premature failures; that is, they increase the speed at which the system ages. Usually, such violations affect only certain used exemplars of the system.

<u>Aging (wear and tear of a system).</u> No matter how good the quality of the unit and the system as a whole, a gradual aging (wear) is inevitable. During the course of use and storage, irreversible changes take place in metals, plastics, and other materials and the accumulative effect of these changes destroys the strength, coordination, and interaction of the parts and, in the final analysis, causes failures. Thus, variations in the lifetime are caused by variations in the quality of the manufacture, the conditions of use, and aging process."

If this book triggers efforts for a clearer justification, augmentation, or replacement of the safety factor by a "better" methodology, its goals will be amply fulfilled.

Indeed, one of the pioneers of the probabilistic analysis of safety factors, Streletskii (1947) notes:

"The concept of the safety factor is directly connected with the security and efficiency of our structures. Despite this fact, one cannot state that this concept was deciphered. Contrary, from all the questions of the analysis of structures the question of the safety factor is most intuitive. Therefore it is of certain interest to attempt to provide an analytical basis for it."

Since 1947 numerous studies have been conducted to achieve this goal. This book describes some of these developments and, hopefully, provides some novel ideas.

Before we proceed, we ought to reply to a question that a thinking reader may ask: Who was the first investigator who introduced the safety factor? Some investigators ascribe this to Coulomb (Randall, 1973). Bernshtein (1957) states that this concept is due to Louis Marie Henri Navier, who also introduced the very concept of stress. He introduced the idea of design according to *working stress* or *allowable stress*, which is obtained by dividing the limiting stress by the factor of safety. Bernshtein (1957) writes (p.49):

"If Galilei was the founder of the science of strength, then Navier was the one who was able to connect it with life: Hence the year 1826 -- the date of publication of the book by Navier-- is not less important in history of this science, the year 1638--the date of its conception."

This is how Bernshtein describes the event of publication of Galileo Galilei's (1638) book:

"In 1638 in a private villa in Arcerti outside Florence, where his last days were lived by Galilei, a book was brought from a faraway Dutch city of Leiden. The book was first printed in an Italian language in the publishing firm Elsevier. The book's title was "Dialogues Concerning Two New Sciences." It was necessary to search for a publisher all over Europe, who would agree to publish a new book of the scientist who was condemned by inquisition. Seventy four years old Galilei took a book, checked it by hands, and put it aside: already a year had passed, since he became blind. In this book of Galilei the foundations were laid for "two new fields of sciences": dynamics and the theory of strength."

Galilei considered the strength of the beam in 1638; however it took 188 years, until in 1826 Navier provided a rigorous solution. Reader can consult the section 2.6 for the more detailed account on the priority question (the reader may read short biographies of Coulomb and Navier in the Appendix B).

Prior to completing these preliminary remarks, one has to answer the following nagging question: "Why is the title of the book posed as a dilemma? Why should safety factor and reliability be either friends or foes?" Perhaps they are neither, i.e. totally unrelated. Perhaps they are both; in other words, may be they have the "love-hate" relationship encompassing elements of both intricate comradery and enmity. Fischer (1970) in his book on the historians' fallacies describes the so called "fallacy of false dichotomous questions." He notes, after giving some examples:

"Many of these questions are unsatisfactory in several ways, at once. Some are grossly anachronistic; others encourage simple-minded moralizing. Most of them are shallow. But all are structurally deficient in that they suggest a false dichotomy between two terms that are neither mutually exclusive nor collectively exhaustive."

Still, it appears that the title of this book, as well as a title of a definitive study by Ditlevsen "Uncertainty and Structural Reliability: Hocus Pocus or Objective Modeling",

10

(1988) where the title pinpoints to a dilemma can be amply defended. It appears to us fully legitimate to raise almost any question, even if it may bear a provocative character. The answer to it may turn out to be that the question itself is not fully valid. Thus, we exercise a more tolerant approach than that adopted by Fischer (1970). For example, title of the paper by Dresden (1992), "Chaos: A New Scientific Paradigm or Science by Public Relations" appears to this writer quite timely and appropriate.

Author is grateful to Dr. Vladimir D. Raizer of FC & T Corporation, Professor Niels Lind of the University of Waterloo, Canada, Professor Michael Hasofer of the University of Victoria, Australia, and Professor Marco Savoia of the University of Bologna, Italy for providing constructive comments (the latter magnanimously checking the entire manuscript and providing with numerous and most helpful suggestions), Dr. Giora Maymon of Rafael, Israel for allowing the use of figures from his papers (2002), Professor Benjamin Reiser of the Haifa University, Israel for providing some of the references, and Dr. Dimitri Val of the Technion-Israel Institute of Technology for help in locating some portraits, and the works by late Professor Streletskii. I am indebted to Professor Victor Birman of the University of St. Louis, and late Professor Joseph Kogan of the Technion-Israel Institute of Technology, for making available some books on structural reliability in the Russian language, some decades ago. Grateful thanks are to Professor V. P. Chirkov of the Moscow Power Engineering Institute and State University, Russia; Dr. Moshe Danieli of the Ariel Center, Israel; Professor Sandor Kaliszky and Professor Péter Lenkei of the University of Budapest, Hungary; Dr. Haim Michlin and Ing. Eliezer Goldberg of the Technion-Israel Institute of Technology; Professor Sergo Evsadze of the Georgian Polytechnic University, Tbilisi; Professor Lia Mukhadze of the Institue of Structural Mechanics and Earthquake Engineering of the Georgian Academy of Sciences, Tbilisi, Georgia; Professor Peter Zimmermann of Universität des Bundeswehr, Federal Republic of Germany; Professor Robert Heller of the Virginia Polytechnic Institute and State University; Professor Lothar Gaul of the Universität of Stuttgart, Federal Republic of Germany; and last but not least, Professor Theodore I. Bieber of the Florida Atlantic University for their kind help in locating various biographical materials. Sincere thanks are expressed to Professor C.W. Bert of the University of Oklahoma for communicating over 20 references in the final stages of printing the book.

Grateful thanks are expressed to the NASA Glenn Research Center (Program Monitor: Dr. Christos C. Chamis) and ICASE-NASA Universities Space Research Association (Program Director: Dr. W. Jefferson Straud) for financial support that led to the investigation conducted in this book. An international team of helpers must be gratefully thanked for skillful typing. These are Mr. Denis Meyer of Paris, France; Mrs. Margaret Pettersdottir of Reykiavik, Iceland; Ms. Giulia Catellani of Modena, Italy; Mr. Silvio Lacquaniti of Messina, Italy; Ms. Roberta Santoro of Palermo, Italy; Ing. Pablo Vittori of Buenos Aires, Argentina and Mrs. Trudy Jeffries, Ms Allison Marshall, and Mr. James Endres of the U.S. Ms. Cristina Gentilini of Bologna, Italy has been the godsend for providing numerous updated versions of the manuscript. I do not know how to thank them.

Last but not least, the author is grateful to his students, Mr. Michel Fuchs,

Ms. Giulia Catellani, Mr. Venkataswamy
Timothy Masters, Mrs. Leni R. de Morais, ...
Pablo Vittori of the graduate course "Safety ...shnan, Ms. Allison Marshall, Mr.
such a course in the world), given during the...n, Mr. Zachary Suttin and Ing.
2002/2003 at the Florida Atlantic University. Tr...eliability" (possibly the first
various portions, provided insightful comments and p...ster of the academic year
first presentation of this material was given at the short ...ingly checked the text's
of Catania, Italy, in summer 2000. ...ome calculations. Very

Author will be most appreciative to hear from ...urse at the University
elishako@fau.edu, for communications about possible impr...aders, by e-mail
information on additional material that ought to be included, prac...ts to the text,
successful implementation of the interplay between the safety factors ...ases of design,
serious or humorous comments on failure, probability, reliability, safety...iability, and
concepts, for future possible edition(s). If included, readers' contributi...associated
acknowledged. ...vill be

Ms. Giulia Catellani, Mr. Venkataswamy Jayalabalakrishnan, Ms. Allison Marshall, Mr. Timothy Masters, Mrs. Leni R. de Morais, Mr. John Simon, Mr. Zachary Suttin and Ing. Pablo Vittori of the graduate course "Safety Factors and Reliability" (possibly the first such a course in the world), given during the Fall Semester of the academic year 2002/2003 at the Florida Atlantic University. They painstakingly checked the text's various portions, provided insightful comments and performed some calculations. Very first presentation of this material was given at the short graduate course at the University of Catania, Italy, in summer 2000.

Author will be most appreciative to hear from the readers, by e-mail elishako@fau.edu, for communications about possible improvements to the text, information on additional material that ought to be included, practical cases of design, successful implementation of the interplay between the safety factors and reliability, and serious or humorous comments on failure, probability, reliability, safety and associated concepts, for future possible edition(s). If included, readers' contributions will be acknowledged.

Chapter 2
Reliability of Structures

"...It seems absurd to strive for more and more refinement of methods of stress-analysis if in order to determine the dimensions of the structural elements. Its results are subsequently compared with so called working stress, derived in a rather crude manner by dividing the values of somewhat dubious material parameters in conventional materials tests by still more dubious empirical numbers called safety factors."

A.M. Freudenthal

"Reliability is a most important property of a mechanical system... Reliability is a useful mathematical concept in striving to provide society with optimal mechanical system."

N. Lind

"All quantities (except physical and mathematical constants) that currently enter into engineering calculations are in reality associated with some uncertainty. If this were not the case, a "safety factor" only slightly in excess of unity would suffice in all circumstances."

P. Thoft-Christensen and M.J. Baker (1982)

"Despite what we often think, the parameters of the loading and the load-carrying capacities of structural members are not deterministic quantities (i.e., quantities which are perfectly known). They are random variables, and thus absolute safety (or zero probability of failure) cannot be achieved. Consequently, structures must be designed to serve their function with a finite probability of failure."

A. S. Nowak and K.R. Collins (2000)

This chapter represents a brief review of the concepts of the reliability of structures. Some closed form solutions, as well as approximate methods are elucidated. Different attempts to describe probabilistically the so-called "safety factor" are provided as well as and some instructive counter-examples are given.

Die
Sicherheit der Bauwerke

und ihre Berechnung nach Grenzkräften anstatt nach zulässigen Spanungen

Von

Dr.-Ing. Max Mayer

Duisburg

Mit 3 Textabbildungen

Berlin

Verlag von Julius Springer

1926

The title page of Max Mayer's book "Safety of Building Structures", published in 1926 in Berlin, in the German language

Prof. Alfred Martin Freudenthal
(born in Poland, February 12, 1906, died in USA, September 27, 1977)

המתחים המותרים ובטחון המבנים

מעטים הם המושגים התכניים הנראים לו למהנדס כה ראויים לאמון וכה חד⁻משמעיים כמושג ה„מתחים המותרים". אולם למעשה לא מרובים הם המושגים התכניים הברורים פחות ממושג זה. מה משמעו של מושג זה? כלום המתח המותר סגולת החומר הוא? או קים קשר ישר בינו ובין סגולה כזאת ומהו אותו קשר? מה היא סגולת החומר הקובעת אותו ואיך היא ניתנת להקבע? כלום העלאת המתח המותר משמעה הפחתת הבטחון ובאיזו מידה? כל אלה הן שאלות אשר התשובות להן קובעות את מהותו של מושג „המתח המותר". את בהינת המושג הזה יש להתחיל מהקשר המהותי ביותר, כלומר מהקשר אשר בין המתח המותר והבטחון. רק נתוח מוקדם של מושג הבטחון עלול לתת לנו את האפשרות להוציא משפט בדבר ערכו או חוסר⁻ערכו של מושג המתח המותר.

כאן עלינו לצאת מתוך הכרת העובדה היסודית שבגלל הפגמים בתפיסת⁻החומר ובתורות⁻ החוזק ומפאת חוסר הבטחון שבתהליכי⁻היצור ובמדירות ראשית המעמסה (מצב המא⁻ מצים) המחושבת של המבנים להוות אך חלק מהמעמסה (מצב המאמצים) אשר המבנים האלה יכולים היו לעמוד בפניה לפי לקח הנסיון המעשי. את מנת המעמסה למעשה ממעמסת⁻השבר נוהגים לכנות בשם מנת הבטחון או הבטחון סתם. על מנת⁻הבטחון הזאת הוטל למלאות ולתקן את אי⁻הדיוקים דלקמן:

1. סטיות ופגמים באחידות בייצור חמרי הבנין;
2. אי⁻דיוקים וחוסר⁻ודאות בקביעת סגולות החוזק;
3. סטיות מותרות במדות והפרשי⁻משקל;
4. השפעות שאינן ניתנות להקבע בדרכי הטפול והעבוד על הבנין, מתחי⁻ייצור נוספים וכו';
5. שנויים הבאים עם השמוש כנון החלדה, בלאי וכו';
6. פגמי חוסר⁻ודאות וחוסר⁻בהירות בחשבון לגבי

   א. השפעות חוץ (סטטיקה) וכמו כן לגבי

   ב. כחות פגימיים (תורת התנגדות החמרים).

מהאמור לעיל מסתבר שלא החומר עצמו קובע את מדת הבטחון אלא התנהגותו בבנין ומאידך גיסא לעולם יש לבסס את הבטחון על השבר או על מצב בלתי⁻רצוי של המבנה ולא על סגולת⁻החומר שהשניחו בה בלבד, סגולה הנקבעת לפי שיטה הנתונה פחות או יותר לשרירות⁻לב ואשר לעתים תכופות איננה נותנת את ערכה האמתי של התכונה כי אם מספר המשמש להשואה נרידא. הדוגמה הידועה ביותר הוא הערך הנקבע מתוך נסיון דחיסת הקביות כמדת התנגדות הבטון ללחץ והנחשב כשוה לה, בה בשעה שהגודל הנקבע באופן זה זה איננו מזהה, ולו גם בקירוב, עם ההתנגדות הפיסיקלית ונמצא בתלות אופינית בשיטת הנסיון.

חוץ מהההשפעות שנזכרו לעיל יש להביא לידי הבעה במנת⁻הבטחון עוד מסבה חשובה אחת, והיא מהות המבנה עצמו. משמעו של הדבר שבעית הבטחון איננה טכנית בלבד כי אם

First title page of the first paper entitled "Allowable Stresses and Safety of Structures" by A. M. Freudenthal in probabilistic mechanics, published in 1938, (in Hebrew)

Prof. Aleksei Rufovich Rzhanitsyn (1911-1987)

2.1   Introductory Comments

The first harbinger of the new discipline of "probabilistic mechanics" appeared in Germany, in 1926, in a book by Mayer (1926). This method was then developed in Russia by Khozialov (1929), Streletskii (1947), Rzhanitsyn (1954, 1959, 1978), in Hungary by Kazinczy (1928, 1952), in England by Tye (1944) and Pugsley (1966), in France by Levi (1949), in Sweden by Johnson (1953), in Georgia by Mukhadze and Kakushade (1954) and fully flourished due to the efforts by Freudenthal, first in Israel and then in the United States (1947, 1956) as well as by many other researchers around the world who followed (short biographical notes on Mayer, Streletskii, Freudenthal, Rzhanitsyn, Mukhadze and Kakushade are summarized in the Appendix B). Now that subject has passed the age of adolescence, as Cornell (1981) suggests, and became a widely accepted discipline with many monographs written and a number of periodical journals appearing, along with numerous regional and international conferences with attendant heavy volumes of proceedings. In the following sections, we will give a brief overview of the basics of the probabilistic approach to structures. The pertinent but not exhaustive references are given in the bibliography.

Prof. Nikolai Stanislavovich Streletskii (born on September 14, 1885; died on February 15, 1967)

ПРЕДИСЛОВИЕ

Работа имеет целью дать, указания по методу определения коэффициента запаса сооружений, исходя из принципа равно-прочности, на основе статистического учета обстоятельств работы сооружения и свойств материала. Метод основывается на некоторых положениях теории вероятностей и приемах статистики, которые могут быть неизвестны инженеру, поэтому в работе приведены о них лишь краткие практический сведения. Несмотря на недостаток опытного материала, рассматриваемый метод уже сейчас раскрывает картину явления и углубляет природу основных положений нашего расчета, чем объясняется усиливающийся интерес к этому методу со стороны инженеров.

Работа составлена в 1945 г., до появления постановления Совета министров от 13/XII 1946 г. № 2678, согласно которому регламентирован переход на расчет по предельным усилиям. Однако § § 12 и 14, в которых используется методика допускаемых напряжений, оставлены в работе ввиду их прин-ципиального интереса.

1. УСЛОВИЕ НЕРАЗРУШИМОСТИ И ДВЕ КОНЦЕПЦИИ РАСЧЕТА

Основным принципом инженерного дела и инженерного расчета является условие неразрушимости, согласно которому действующее в сооружении (конструкции) наибольшее усилие за время со службы должно быть меньше пли, в крайнем случае, равно наименьшему возможному па эго время предельному сопротивлению материала конструкции

$$\text{действ.}\ \ : \ \ \text{'Алт. и }><)^{\wedge\wedge}\text{пин} \atop \text{констр}\ \ \text{о т/т. кометы}^{\bullet} \tag{1.1}$$

В соответствии с этим основным вопросом инженерного расчета является определение этих усилий.

Несомненно, задача эта является исключительно сложной, так как мы в ней имеем дело с гипотетическими усилиями, которые мы можем предвидеть только с определенной долей вероятия. Однако, идя статистическим путем, изучая и сопоставляя факты работы однородной группы сооружений и материала в конструкциях, мы можем установить закон появлении этих фактов и экстраполировать этот закон па будущее, если будем иметь к тому достаточные основания.

По этому закону может быть получено и предельное усилие, если мы его подчиним некоторым наперед заданным условиям (например сроку службы сооружения, вероятию появления предельного усилия и т. д.). Зная предельное усилие и разделив его на коэфициент, который обычно у нас носит название коэфициента запаса, мы получаем расчетное усилие

$$_р \qquad \text{афопт. пред}^{\text{лик}}$$

Таким образом расчетное усилие есть функция предельного усилия и выбранного коэфициента запаса. Отсюда следует, что расчетное усилие является усилием условным, мы можем взять меньший коэфициент запаса и получить большее рас-

Н. С. СТРЕЛЕЦКИЙ

ИЗБРАННЫЕ
ТРУДЫ

Под редакцией засл. деят. науки и техники РСФСР
проф. д-ра техн. наук Е. И. Беленя

The title page of Prof. N.S.Streletskii's book "Selected Works." His book "Statistical Basis of the Safety Factor of Structures " was published in 1947 in the Russian language.

2.2 Basic Concepts

Consider the situation where the state of a structure in use can be described by a finite number of probabilistically dependent or independent parameters X_1, X_2. ..., X_n, part of which characterizes the loadings acting on the structure, and the other part is associated with the strength of the materials. For some combinations of its parameters the system is "acceptable" for use (in which case it is said to be in the safe state), whereas for other combinations it is "unacceptable" (in the failed state). The function $f(x_1, x_2, ..., x_n)$, x_j being the possible value of the random variable X_j may take on, which vanishes at the transition surface between the two states, is so defined that its positive values

$$f(x_1, x_2, ..., x_n) > 0 \qquad (2.1)$$

represent the safe state, while its negative values

$$f(x_1, x_2, ..., x_n) < 0 \qquad (2.2)$$

represent the failed state.

For example, if parameters X_1, X_2, ... , X_m represent the strengths of the materials or capacities denoted by C, and X_{m+1}, X_{m+2}, ..., X_n - the actual stresses denoted by S, the failure surface could be put in the form

$$f = C(x_1, x_2, ..., x_m) - S(x_{m+1}, ..., x_n)$$

$$\equiv M(x_1, x_2, ..., x_n), \qquad (2.3)$$

with $M(x_1, x_2, ..., x_n)$ representing the safety margin.

The reliability of the structure - the probability of its being in the safe state is obtained as

$$R = Prob(f > 0) = \int_0^\infty f_M(m)\,dm \qquad (2.4)$$

The formal definition of reliability reads, according to the McGraw-Hill Dictionary of Scientific and Technical Terms (1975):

> "Reliability: the probability that a component part, equipment, or system will satisfactorily perform its intended function under given circumstances, such as environmental conditions, limitations as to operating time, and frequency and thoroughness of maintenance, for a specified period of time."

The probability of failure or the unreliability of the structures reads

$$P_f = \int_{-\infty}^0 f_M(m)\,dm, \qquad (2.5)$$

In Eqs. (2.4) and (2.5) $f_M(m)$ is the probability density of the safety margin which can be found through the familiar expression for the probability density of the difference of the random variables C, S. Referring to the concept of failure, Gertsbakh (1989) writes:

> "The word "reliability" refers to the ability of a system to perform its stated purpose adequately for a specified period of time under the operational conditions encountered. Any system will be absolutely reliable of some undesirable events, called failures, do not occur in the system's operation. Every system has its own set of such undersirable events. For example, a failure of a watch may be defined as a delay exceeding 5 s over a 24-h period. For a mechanical system, a failure is a breakdown (a crack) of some of its parts or an increase in vibration above the permitted level, etc. One of the most dangerous failures of a nuclear reactor is a leak of a radioactive material. For a missile, the failure could mean missing the target or exploding before hitting it. A military aircraft servicing system fails if it is not able to provide aircraft availability above some prescribed level, say 0.95."

Fig. 2.1 depicts the probability of failure P_f and the probability of success P_s that equals reliability R.

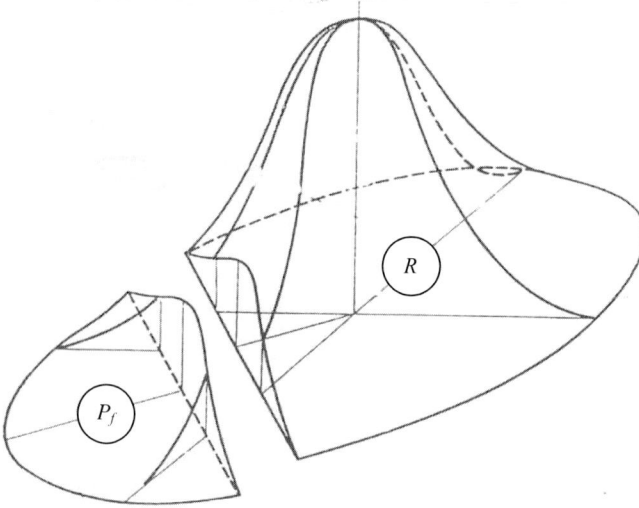

Fig. 2.1 Probability of failure and probability of success. The latter equals the reliability R

One may ask a legitimate question: Why not to demand absolute 100% reliability? Thoft-Christensen and Baker (1982) as it were, answer this inquiry; they stress that "the lack of information about the actual behavior of structures combined with the use of codes embracing relatively high safety factors can lead to the view, still held by some engineers as well as by some members of the general public, that absolute safety can be achieved. Absolute safety is of course unobtainable; and such a goal is also undesirable, since absolute safety could be achieved only by deploying infinite resources. It is now widely recognized, however, that some risk of unacceptable structural performance must be tolerated." An immediate question follows this discussion: How much risk is being tolerated by the public?

Thoft-Christensen and Baker (1982) note:

> "For example, although air travel is associated with a high risk per hour, a typical passenger may be exposed for between only, say 10-100 hours per year, leading to a risk of death of between 10^{-5} and 10^{-4} per year (i.e. between 1 in 10^5 and 1 in 10^4). In contrast, most people spend at least 70% of their life indoors and are therefore exposed to the possible effects of structural failure, but this leads to an average annual risk per person of only 10^{-7}." The Central Statistical Office of the United Kingdom reported that in the years 1970-1973, the number of deaths per hour per 10^8 persons was 2700 due to mountaineering, 120 due to air travel, 59 due to deep water trawling, 56 due to car travel, 21 due to coal mining, 7.7 at the construction sites, 0.1 due to a fire at home, whereas it constitutes 0.002 for the structural failure."

Since the society *de facto* accepts some risk, the probabilistic analysis of structures is far from being nonsensical. Madsen, Krenk and Lind (1986) note:

> "Allen (1981) estimated that the total number of structures (5 million in Canada) in service and the number of failures per year (also in Canada). This gives a failure rate of 2×10^{-5} per year from all causes. Human error is estimated to account approximately 90% of these failures, leaving a design reliability of "error free" structures of 2×10^{-6} per year or 10^{-4} for a 50-year service life."

For the probability distribution function of the safety margin M, we get

$$F_M(m) = \int_{-\infty}^{\infty} f_C(s+c) f_S(s)\, ds, \qquad (2.6)$$

where $f_C(c)$ is the probability density of the capacity and $f_S(s)$ - that of the stress. Irrespective of the specific densities of C and S, we have for the safety margin the mean

$$E(M) = E(C) - E(S) \qquad (2.7)$$

For the uncorrelated C and S we get the following expression of the variance

$$Var(M) = Var(C) + Var(S). \qquad (2.8)$$

The number of standard deviations of the safety margin in the interval $m = 0$ to $m = E(M)$ is called reliability index (Fig. 2.2):

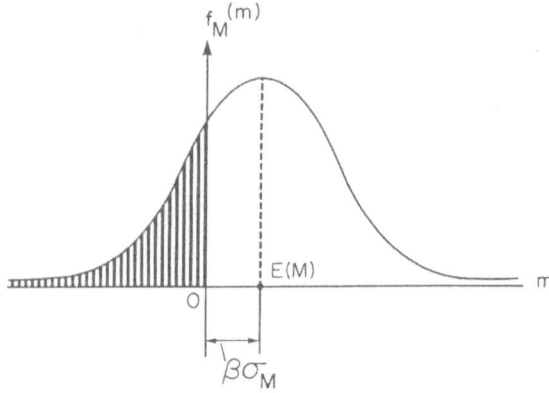

Fig. 2.2 Probability density of the safety margin

$$\beta = \frac{E(M)}{\sigma_M},$$
(2.9)

where $\sigma_M = \sqrt{Var(M)}$ is the mean square deviation of the safety margin.

For the case where C and S are correlated, we have instead of Eq. (2.9),

$$\beta = \frac{E(C) - E(S)}{\left[Var(C) - 2Cov(C, S) + Var(S)\right]^{1/2}}.$$
(2.10)

If C and S are normally distributed, then the reliability and probability of failure equal, respectively,

$$R = \Phi(\beta),$$
(2.11)

$$P_f = \Phi(-\beta),$$
(2.12)

where $\Phi(x)$ is the normal cumulative distribution function

$$\Phi(x) = \frac{1}{\sqrt{2\pi}} \int_{-\infty}^{x} e^{-t^2/2} dt.$$
(2.13)

Fig. 2.3 shows the iso-probability curves, ellipses for the general case of $\sigma_R \neq \sigma_S$.

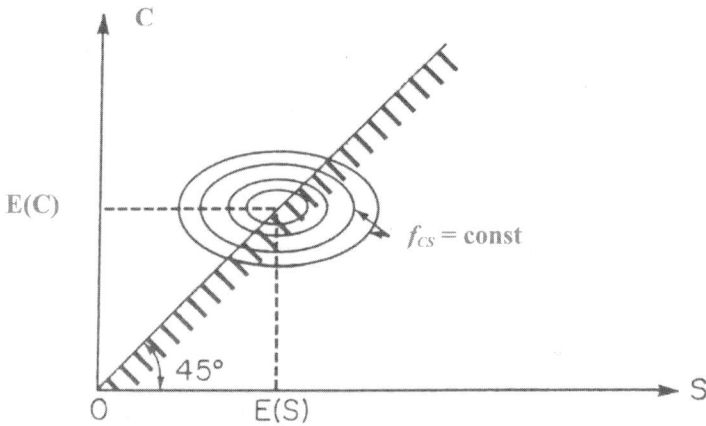

Fig. 2.3 Curves of equal joint probability density of stress and capacity (strength)

The reliability index β has an interesting geometrical interpretation using the standard independent normal variables, when the iso-probability curves become circles (Fig. 2.4):

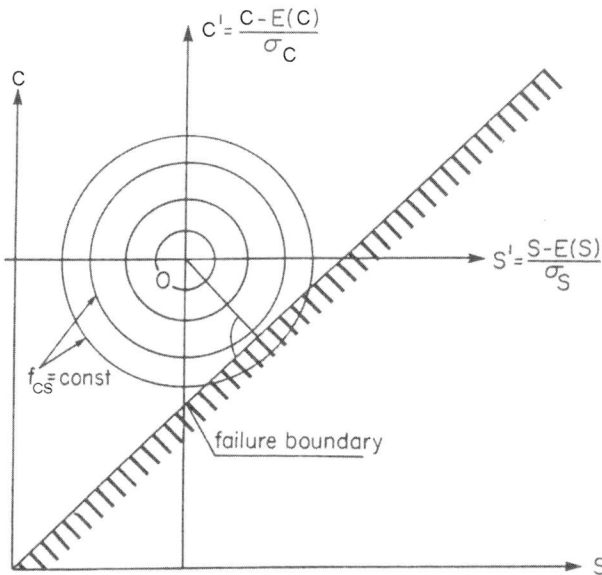

Fig. 2.4 Curves of equal probability density in the standard space and the failure boundary

$$C' = \frac{C - E(C)}{\sigma_c}, \quad S' = \frac{S - E(S)}{\sigma_s}. \tag{2.14}$$

The failure surface (2.3) is rewritten as

$$\sigma_c C' - \sigma_s S' + [E(C) - E(S)] = 0. \tag{2.15}$$

According to the analytical-geometry formula, the distance from the origin to the failure surface is

$$d = \frac{E(C) - E(S)}{\sqrt{\sigma_c^2 + \sigma_s^2}}, \tag{2.16}$$

which is formally identical with the expression for the reliability index β in Eq. (2.10) (Fig. 2.5).

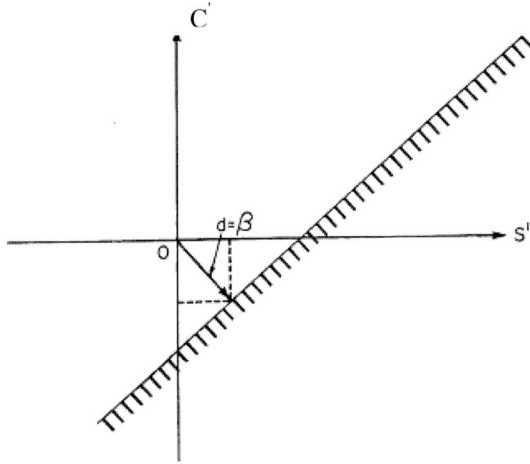

Fig. 2.5 Interpretation of minimum distance for the linear failure boundary

Point where d meets the failure surface is called the design point. Thus, we arrive at:

$$R = \Phi(\beta) = \Phi(d), \tag{2.17}$$

$$P_f = \Phi(-\beta) = \Phi(-d) \tag{2.18}$$

There are few other situations in which exact expressions can be obtained for the reliability. Let, for example, the strength and stresses be independent, and the marginal densities exponential:

$$f_C(c) = \frac{1}{E(C)} \exp\left[-\frac{c}{E(C)}\right], \tag{2.19}$$

$$f_S(s) = \frac{1}{E(S)} \exp\left[-\frac{s}{E(S)}\right], \tag{2.20}$$

where $E(C)$ and $E(S)$ are the mean strength and stress, respectively. We obtain the following expression for the reliability:

$$R = \frac{E(C)}{E(C) + E(S)}.$$ (2.21)

Analogously, if C and S are independent random variables, having Rayleigh distribution

$$f_C(c) = \frac{\pi r}{2[E(C)]^2} \exp\left\{ -\frac{\pi c^2}{4[E(C)]^2} \right\},$$ (2.22)

$$f_S(s) = \frac{\pi s}{2[E(S)]^2} \exp\left\{ \frac{\pi s^2}{4[E(S)]^2} \right\},$$ (2.23)

the reliability becomes

$$R = \frac{[E(C)]^2}{[E(C)]^2 + [E(S)]^2}.$$ (2.24)

Additional important case is when both the stress and the strength have a log-normal distribution:

$$f_S(s) = \frac{1}{s\sigma_1\sqrt{2\pi}} \exp\left[-\frac{(\ln s - a)^2}{2\sigma_1^2} \right], (s \geq 0)$$ (2.25)

$$f_C(c) = \frac{1}{c\sigma_2\sqrt{2\pi}} \exp\left[\frac{(\ln c - b)^2}{2\sigma_2^2} \right], (c \geq 0)$$ (2.26)

where a, b, σ_1 and σ_2 are the density parameters, so that

$$E(S) = \exp\left(a + \frac{1}{2}\sigma_1^2 \right),$$

$$E(C) = \exp\left(b + \frac{1}{2}\sigma_{21}^2 \right),$$ (2.27)

$$Var(S) = \exp\left(2a + \sigma_1^2\right)\left[\exp\left(\sigma_1^2\right) - 1\right],$$

$$Var(C) = \exp\left(2b + \sigma_2^2\right)\left[\exp\left(\sigma_2^2\right) - 1\right].$$

The reliability is then

$$R = Prob\left(V = \frac{S}{C} \leq 1 \right) = F_V(1),$$ (2.28)

where $F_V(v)$ is the probability distribution of the random variable V. Eq. (2.28) may be written as

$$R = Prob(\ln V \leq 0) = F_{\ln V}(0).$$ (2.29)

26

Prof. Abraham Michael Hasofer
(born on October 2 1927 in Alexandria, Egypt)

Note that

$$\ln V = \ln S - \ln C, \qquad (2.30)$$

and since $\ln Sl$ and $\ln C$ both have a normal distribution, specifically $\ln S$ is $N(a,\sigma_1^2)$ and $\ln C$ is $N(b,\sigma_2^2)$, $\ln V$ is also normal as a difference of normal variables $N(a-b, \sigma_1^2 + \sigma_2^2)$, implying that V is log-normal.

Reliability becomes

$$R = \Phi\left(-\frac{a-b}{\sqrt{\sigma_1^2 + \sigma_2^2}}\right). \qquad (2.31)$$

For other cases where exact solutions are obtainable, one should consult with the monographs by Ferry Borges and Castanheta (1971), Rzhanitsyn (1954), Bolotin (1981), Ang and Tang (1984), Augusti, Baratta and Casciati (1984), Elishakoff (1983,1999), Ditlevsen (1981), and others.

How does one calculate the reliability of a structure where exact solutions are unavailable? Such is usually the case if the relationship (2.1) is nonlinear. Under these circumstances, if the basic variables are still normally distributed, the formula (2.17) and (2.18) are still used, but now as approximations to the exact reliability and probability of failure, respectively (Fig. 2.6).

Prof. Niels Lind
(born in Copenhagen, Denmark, March 10, 1930)

Equations (2.11) and (2.12), when the failure boundary is nonlinear, are commonly referred to as "Hasofer and Lind" index (1974) on account of their systematic developments which lead to the wide-spread characterization of the reliability index as the minimal distance from the origin to the nonlinear failure surface. It appears instructive to give a quote from the paper by Shinozuka (1983):

> "...it is worth noting that the checking format for a modified design, recommended on the basis of these recent developments was in essence suggested by Freudenthal (1956). In this paper, referring to what is now known as the checking point, we wrote " because the critical conditions (x^0, y^0) has the highest probability of occurrence along the line $r = 0$, it represents the combination to be used in design." The critical failure condition (x^0, y^0) indicates the point on the limit state (or failure) surface $f = 0$ and located at the shortest distance, ρ, from the origin on the two-dimensional rectangular Cartesian coordinate space of the standardized Gaussian variables x an y. The critical point is also the point of maximum likelihood due to the Gaussian property assumed in the design variables. L.S. Lawrence (see Freudenthal, 1956) and J.M. Corso (see Freudenthal, 1956) in their discussion of the Freudenthal paper, pointed out, and Freudenthal concurred, that the limit state probability (probability of failure) is a function of the shortest distance, ρ. This is now known as the safety index, and can be obtained as $\Phi(-\rho)$, where $\Phi(\cdot)$ = the standardized Gaussian distribution function ... his paper did suggest that the checking point used for design and that the checking point of shortest distance from the origin in a standardized Gaussian or transformed Gaussian variable space and at the same time is the point of maximum likelihood."

28

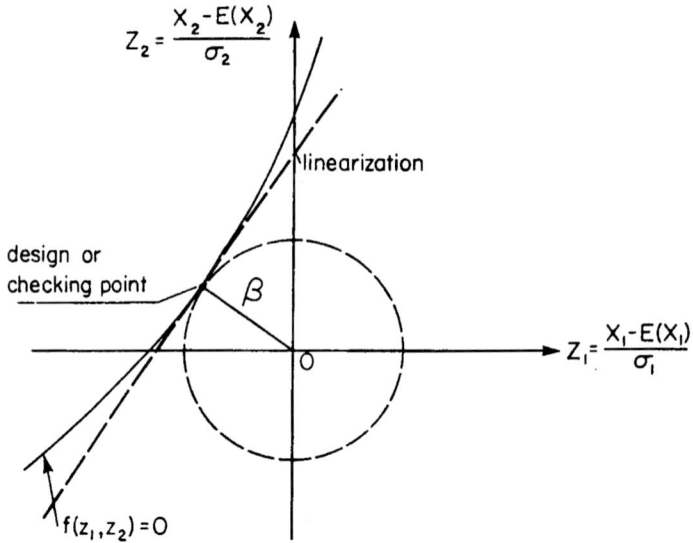

Fig. 2.6 Minimum-distance as the reliability index

Interestingly enough, the description of the minimum distance method appears in the monograph by Olszak *et al* (1961). Murzewski (1989) is attributing this method to Levi (1949). Nevertheless, the seminal paper by Hasofer and Lind (1974) determined, for years to come, the research in structural reliability.

2.3   How Accurate Is Minimum Distance Reliability Index?

Schuëller (1993) notes,

> "Approximate methods…may, for some cases, provide inaccurate results. Unfortunately, they still lack mathematically based guidelines concerning their range of applicability. In other words, for all new types of problems, the results still have to be verified by simulation technique. This may be considered as a considerable drawback, particularly for practical application."

It makes sense, therefore, to compare the approximate methods to the exact solutions, to have more insight, even if for simple problems only. Comparison with simulation technique, mentioned in the above quote of Schuëller (1993) appears to be better left to the complex structural configuration, or the ones that are not amenable to the exact evaluation.

To best answer the question that was posed is the title of this section, we will study the circular shaft-case amenable to exact solution (Elishakoff, 1987). Given a circular shaft

(Fig. 2.7) of radius α subjected simultaneously to a bending moment M and torque T, characterized as random variables with joint probability density function $f_{MT}(m,t)$; the yield stress σ_y is constant with probability unity, or, in other words, deterministic.

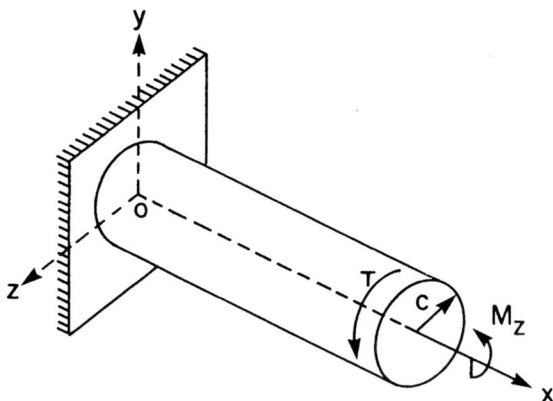

Fig. 2.7 Circular shaft subjected to random bending moment and torque

According to the maximum theory of failure, the stress theory of failure, the strength requirement reads

$$\frac{M_{eq}\,\alpha}{I} \leq \sigma_y \tag{2.32}$$

where M_{eq} is the "equivalent" moment, and I is the moment of inertia of a circular area of radius c:

$$M_{eq} = \sqrt{M^2 + T^2}, I = \frac{\pi\alpha^4}{4} \tag{2.33}$$

and reliability becomes

$$R = Prob\left(\sqrt{M^2 + T^2} \leq \frac{\pi}{4}\sigma_y\,\alpha^3\right). \tag{2.34}$$

Consider first the simplest case treated by Bolotin (1971): M and T are independent normal variables with zero means ($a=b=0$) and equal variances $\sigma_M = \sigma_T = \sigma$. Then the equivalent moment has a Rayleigh distribution

$$f_{M_{eq}}(m_{eq}) = \frac{m_{eq}}{\sigma^2}\exp\left(-\frac{m^2_{eq}}{2\sigma^2}\right), \tag{2.35}$$

with the attendant reliability

$$R = 1 - \exp\left(-\frac{\pi^2\sigma_y^2\alpha^6}{32\sigma^2}\right) \tag{2.36}$$

To compare this exact expression with the minimum distance method, we introduce the basic variables

Prof. G. Mukhadze (born on November 29, 1886 in Tbilisi, Georgia; died on January 5, 1963 in Tbilisi, Georgia)

$$Z_1 = \frac{M}{\sigma}, \qquad Z_2 = \frac{T}{\sigma} \tag{2.37}$$

The failure boundary becomes

$$Z_1^2 + Z_2^2 \leq \rho^2 \tag{2.38}$$

where

$$\rho = \frac{\pi}{4}\frac{\sigma_y}{\sigma}\alpha^3 \tag{2.39}$$

Eq. (2.38) represents a circle with radius ρ. The minimum distance to the circle equals the radius itself, so that $\beta = \rho$. Hence, under the minimum distance approximation we have

$$R = \Phi(\rho), \qquad P_f = \Phi(-\rho) \tag{2.40}$$

whereas the exact solution (Eq. (2.36)) in terms of ρ is

$$R = 1 - \exp\left(-1/2\rho^2\right), \qquad P_f = \exp\left(-1/2\rho^2\right) \tag{2.41}$$

For highly reliable structures, which is where our interest lies, the approximate solution is remarkably close to the exact one.

For example, if $\rho = 3.35$, the exact reliability value is $R=0.9963$ whereas the Hasofer-Lind approximation yields $\overline{R} = 0.9996$. Agreement diverges for smaller values of ρ: For $\rho = 2$, the exact value is $R = 0.86467$, and the approximation constitutes $\overline{R} = 0.97725$.

Consider now the more realistic case $a^2 + b^2 \neq 0$ with the former restriction $\sigma_M = \sigma_T = \sigma$ still retained. Elishakoff (1983, 1999) furnished an exact solution in the series form; here we will provide the closed-form solution. In terms of the basic variables, $Z_1 = (M - a)/\sigma, Z_2 = (T - b)/\sigma$, the probabilistic counterpart of Eq. (2.31) reads:

$$R = Prob\left[A \equiv \left(Z_1 + \frac{a}{\sigma} \right)^2 + \left(Z_2 + \frac{b}{\sigma} \right)^2 \leq \rho^2 \right]. \tag{2.42}$$

The random variable A has a non-central chi-square distribution with two degrees of freedom, (Johnson and Kotz, 1970) and the noncentrality parameter δ and associated reliability R:

$$\delta^2 = \frac{\left(a^2 + b^2 \right)}{\sigma^2}. \tag{2.43}$$

$$R = Prob\left[\chi_2'^2 (\delta^2) \leq \rho^2 \right]$$

the prime denoting 'non-central'. Extensive tables of non-central chi-square distributions are available [see e.g. Haynam, Govindarajulu and Leone (1962)]. For our purposes it is instructive to use an accurate approximation for large values of δ^2, since we assume $a/\sigma \gg 1$ and $b/\sigma \gg 1$. Johnson and Kotz (1970) give the useful results for various ranges of ρ.

Thus, the reliability can be found via the extensive tables available. In the case $a/\sigma \gg 1$, $b/\sigma \gg 1$, an asymptotic expression is available, valid for $\rho > 5$:

$$R = Prob\left(A \leq \rho^2 \right) = Prob\left(\chi_2'^2 (\delta^2) \leq \rho^2 \right) \approx \Phi\left(\sqrt{\rho^2 - 1} - \delta \right) \tag{2.44}$$

Prof. A. Kakushadze (born November 4, 1903 in Patriketi, Georgia; died July 29, 1981 in Tbilisi, Georgia)

Let us compare this result to that obtained by the minimum distance method. The failure boundary is again a circle with radius ρ, but now centered at $(-a/\sigma, -b/\sigma)$. The minimum distance from the coordinate origin to the circle is (Fig. 2.8):

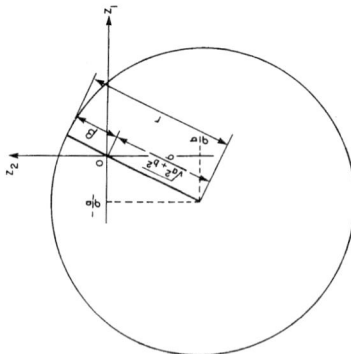

Fig. 2.8 Exact calculation of the minimum distance

$$\beta = \rho - \sqrt{a^2 + b^2} / \sigma = \rho - \delta. \tag{2.45}$$

Hence

$$R \approx \Phi(\rho - \delta), \quad P_f \approx \Phi\left[-(\rho - \delta)\right]. \tag{2.46}$$

Comparison of equations (2.44) and (2.45) suggests that the minimum distance approximation is an excellent one, as for $\rho \gg 1$ the asymptotically exact (Eq. (2.44)) and approximate expression (Eq. (2.46)) tend to each other. Additional results on the accuracy of the Hasofer-Lind method are given in the Appendix A.

2.4   Safety Factors as Discussed in Literature

Numerous attempts at probabilistic interpretation of the safety factor have been made in the literature despite the fact that the "spirit" of these two approaches are different. Before reviewing them, it is instructive to quote the following excerpts from popular textbooks concerning its definition:

a)   "In selecting member sizes and materials the designer must ensure that the *failure load* (the minimum value of the load that, according to specified criteria, would cause failure of the member) is safely above the *allowable load* (the maximum load that the member is allowed to experience during its service lifetime). The *factor of safety*, *FS*, is defined as

$$FS = \frac{failure\ load}{allowable\ load}$$

Of course, *FS*>1. If there is a linear relationship between the loads on the structure and the stresses caused by the loads, it is permissible to define the factor of safety as the ratio of the two stresses, the *failure stress* and the *allowable stress*. For axial deformation, the tensile (or compressive) yield stress σ_y is taken as the stress corresponding to failure by yielding; in direct shear, the shear yield strength τ_y is used"(Craig, 1996).

b)   "In the process of design, we often require that the geometry and materials be so chosen that nowhere does the stress equal the stress yield of the material. This ensures that there is no excessive deflection of the system and further is an assurance that the material does not physically fail (i.e., does not reach the ultimate stress). In any such attempt the designer has certain nagging doubts. Some of these doubts are:

1. How accurate is the constitutive law in representing the particular materials being used?

2. How close to specifications will the properties of the materials be from the manufacturer, and how close will the tolerance be kept in the fabrication of proposed members?

3. How good are the estimations of stress concentrations used in the calculations?

გ. მუხაძე, ა. კაკუშაძე

მარაგის კოეფიციენტების განსაზღვრა მათემატიკური სტატისტიკის მეთოდებით

ტექნიკა და შრომა

19 თბილისი 54

The page of the book "Determination of the Safety Factors by the Methods of Mathematical Statistics, by G. Mukhadze and A. Kakushadze, published in 1954 (in Georgian)

4. Will there possibly greater loads coming onto the system than the loads used in the calculations?

There are many other doubts that may persist in the mind of the designer. There is, in short, a certain *degree of uncertainty* in his or her effort –an unavoidable uncertainty irrespective of the skill and experience of the designer. To decrease this uncertainty the designer uses a stress lower than the yield stress in ascertaining the extreme state of stress permissible. This stress is called the *working stress*, and we may define the *factor of safety n* as

$$n = \frac{yield\ \ stress}{working\ \ stress}$$

The higher the value of n, the more probable it is that the system will not fail during a long enough interval of time. On the other hand, too high a safety factor may make your design noncompetitive. Thus, the assignment of a safety factor in a design depends on many factors one of which is the danger to life and limb of people who are using or who are near the structure. Clearly, we cannot give here any prescription for assigning a safety factor except to say that factors of 2 and 3 are not uncommon for certain endeavors" (Shames, 1989).

c) "Structural failures are always undesirable events. They occur because of ignorance, negligence, greed, or physical barriers; sometimes they are considered an act of God. The probability of failure is often higher for projects involving new materials, technology, and extreme parameters (such as span, height, thickness, and weight) for which there is little or no prior experience. Therefore, the specified design provisions include built-in safety margins: load effects are usually overestimated and resistances are underestimated. However, this safety reserve cannot cover all possible causes of structural failure. From the legalistic point of view, the code defines the acceptable practice. In case of failure, the designer can be found responsible if he or she did not satisfy the code requirements. The acceptable reliability levels implied in current building codes depend on the system of values assigned by the society to human life, material loss, disruption of services, and so on. The actual rate of failure in the building industry can be used as an indication of the importance a society assigns to this sector of its national economy. The number of fatalities and injuries attributed to a structural collapse can be compared with loss of life in car accidents, airplane accidents, and so on. The comparison points out that there is a considerable variation in acceptable failure rate, depending on industry, geographical region, subjective aversion to risk, and tradition" (Nowak, 2000, p.216).

d) "Traditionally, in civil engineering assessments of the risk of failure are made on the basis of allowable *factors of safety*, learned from previous experiences for the considered system on its anticipated environment. Conventionally, the designer forms the ratio of what are assumed to be the nominal values of capacity $\tilde{C}$ and demand $\tilde{D}$, as $FS = \tilde{C} / \tilde{D}$. For example, if the allowable stress is 40,000 lb/in^2 and the maximum calculated stress is 25,000 lb/in^2, the conventional factor of safety would be 1.6. The design is considered satisfactory if the calculated factor of safety is greater than a prescribed minimum value learned from experience with such design. Thus, in concept, in the above example, if a factor of safety of 1.6 were considered intolerable, the system would be redesigned to decrease the maximum induced stress or to increase the strength. In general, the demand function is the resultant of many uncertain components of the system under consideration, such as vehicle loadings, wind loadings, earthquake accelerations, location of the water table, temperatures, quantities of flow, runoff and stress history, to name only a few. Similarly, the capacity function will depend on the variability of material parameters, testing errors, construction procedures, inspection supervision, ambient conditions, and so on. Perhaps of greater importance and common to both are the developed analytical models themselves, and their assumptions (formulas, equations, etc.), that are used to scale parameters. Can one truly expect that these highly empirical formulas, developed decades ago to meet the needs of slide rule oriented practitioner will model the real world with a reliability greater than 99%. To do so would have required careful prior observations of the performance of 100 or more similar structures" (Harr, 1987, p.130).

e)   "The quantity that is traditionally used to maintain proper degree of safety in structural and mechanical design is the *factor of safety*. Generally the factor of safety is understood to be the ratio of the expected strength to the expected load. The strength of the component and the acting load are assumed to be unique in conventional design. However, in practice both the strength and load are variables, the values of which are scattered about their respective expected (or mean) values. This results in an overlap in the distributed values of strength and load that might lead to the failure of the system. In fact, the interference area between the strength and load distributions can be used to compute the probability of failure of the component" (Rao, 1992, p.10).

f)   "A margin of safety must be built into any design to account for uncertainties or a lack of knowledge, lack of control over the environment, and the simplifying assumptions made to obtain results. The measure of this margin of safety is the factor of safety…

There are several issues that must be considered in determining the appropriate factor of safety in design… No single issue dictates the choice of factor of safety.

The value for the factor of safety is a compromise of the various issues, which is arrived at from experience.

Lot of considerations are the primary reason for using a low factor of safety. Large fixed cost could be due to the use of an expensive material, or to using a large quantity of material specified in design to meet a given factor of safety requirement. Weight resulting in higher fuel consumption is an example of higher running costs. In aerospace industry the running costs supersede material costs. Material costs dominate the furniture industry. The automobile industry seeks a compromise between fixed and running costs.

Liability cost considerations push for a greater factor of safety. Though liability is a consideration in all design, the building industry is most conscious of it in determining the factor of safety.

Lack of control or lack of knowledge concerning the operating environment will push for higher factors of safety. Uncertainties in predicting earthquakes, cyclones, or tornadoes will require higher safety factors for design of buildings located in regions prone to these natural calamities. A large scatter in material properties as usually seen in newer materials in an uncertainty that will the use of a larger factor of safety.

Human safety consideration not only push the factor of safety higher but often result in government regulations of the factors of safety such as reflected in building codes.This list of issues affecting the factor of safety is by no means complete, but is an indication of the subjectivity that goes into the choice of the factor of safety. The factor of safety that may be recommended for most applications range from 1.1 to 6" (Vable, 2002).

g) "In the traditional approach to design, the safety factor or margin is made large enough to more than compensate for uncertainties in the values of both the load and the capacity of the system under consideration. Thus, although these uncertainties cause the load and the capacity to be viewed as random variables, the calculations [in the traditional approach] are deterministic, using for the most part, the best estimates of load and capacity. The probabilistic analysis of loads and capacities necessary for estimating reliability clarifies and rationalizes the determination and use of safety factors and margins. This analysis is particularly useful for situations in which no fixed force can be put on the loading, for example, with earthquakes, floods and other natural phenomena, or for situations in which flaws or other shortcomings may result in systems with unusually small capacities" (Lewis, 1987).

For other pertinent quotes on the justification of the safety factors the reader can consult section 3.1.

Most vivid criticism of the traditional safety factor analysis is apparently provided by Freudenthal remarks,

"... it seems absurd to strive for more and more refinement of methods of stress-analysis if in order to determine the dimensions of the structural elements, its results are subsequently compared with so called working stress, derived in a rather crude manner by dividing the values of somewhat dubious material parameters obtained in conventional materials tests by still more dubious empirical numbers called safety factors."

Indeed, it appears to the present author that in addition to its role as a "safety" parameter for the structure, it is intended as "personal insurance" factor of sorts for the design companies.

Probabilistic interpretation of the safety factor is not unique. We will discuss here two of possible approaches towards such an interpretation, with additional discussion given in Section 3.2. The "straightforward" safety factor itself, as the ratio C/S, is a random variable. The question is how to define it in probabilistic terms. A possible answer is the so-called central safety factor:

$$n = \frac{E(C)}{E(S)}, \qquad (2.47)$$

which in certain situations is in direct correspondence with the reliability level. Indeed, for exponentially distributed strength and stress, the reliability, as per Eq. (2.21) reads

$$R = \frac{n}{1+n}, \qquad (2.48)$$

However, to achieve the reliability level of say, 0.999, the required central safety factor should be 999!

The situation is "better" for the case when C and S are independent Rayleigh distributed variables; in terms of the central safety factor c, Eq. (2.24) rewrites as

$$R = \frac{n^2}{1+n^2} \qquad (2.49)$$

Under new circumstances, in order to achieve reliability of 0.999, the central safety factor should be $\sqrt{999} = 31.61$!

For normally distributed strength and stress, the safety index could be written as

$$\beta = d = \frac{n-1}{\sqrt{\gamma_S^2 + n^2 \gamma_C^2}} \qquad (2.50)$$

where $\gamma_S = \sigma_S/E(S)$, $\gamma_C = \sigma_C/E(C)$ are coefficients of variation of the stress and strength, respectively. The central safety factor corresponding to the reliability level r satisfies the quadratic

$$\omega_1 n^2 + \omega_2 n + \omega_3 = 0, \qquad (2.51)$$

where

$$\omega_1 = 1 - \gamma_C^2 \left[\Phi^{-1}(r) \right]^2$$

$$\omega_2 = -2, \qquad (2.52)$$

$$\omega_3 = 1 - \gamma_S^2 \left[\Phi^{-1}(r) \right]^2,$$

where $\Phi^{-1}(\cdot)$ is the inverse of $\Phi(\cdot)$. Under these circumstances, c depends on the reliability level r, but also on the coefficients of variation of the strength and stress.

Analogously, for the stress and strength, which have a log-normal distribution, the central safety factor is given by (Elishakoff, 1983, 1999):

$$n = \frac{\exp(a + \sigma_1^2/2)}{\exp(b + \sigma_2^2/2)}. \qquad (2.53)$$

For $\sigma_1/a \ll 1$ and $\sigma_2 \ll 1$, Leporati (1977) derived the following approximation

$$n = \exp\left\{ \beta \left[\left(\frac{\sigma_1}{a} \right)^2 + \left(\frac{\sigma_2}{b} \right)^2 \right]^{1/2} \right\}.$$ (2.54)

The alternative safety factor is introduced as follows

$$t = E\left(\frac{C}{S} \right)$$ (2.55)

instead of Eq. (2.43). The following counter example is constructed by Elishakoff (1983, 1999 pp. 243-246), in which both C and S have an identical, uniform distribution over interval $[0,\alpha]$. Then on one hand, reliability is just one half, but the factor of safety t turns out to be infinity (!).

One can conclude, that for the reliability calculations one should have information on the required reliability of the structure, with additional parameters $E(C)$, $E(S)$, σ_C, and σ_S specified, not necessarily in their direct connection with the "safety factor." In this connection it is instructive to quote Freudenthal (1972):

> "The predictive use, in structural design and analysis, of the theory of probability implies that the designer, on the basis of his professional competence, is able to draw valid conclusions from the probability figures obtained, so as to justify design decisions which in most cases, hinge on considerations of economy and utility. It is not implied that this use is in itself sufficient to make a design more reliable or more economical, any more than that the avoidance of the probabilistic approach makes it safer."

Also:

> "In fact, an approach based on the direct specification of a very low failure probability alone suffers from a major shortcoming: there is no intrinsic significance to a particular failure probability since no a priori rationalization can be given for the adoption of a specific quantitive probability level in preference to any other, so that the selection of this level remains an arbitrary decision."

2.5   About the Acceptable Probability of Failure

It appears to be instructive to provide a podium for several researchers, on the general nature of the reliability calculations. Augusti, Baratta and Casciati (1984) state:

> "...note that the acceptable probabilities of failure P_f are, in civil engineering, very low: their order of magnitude, as rough quantitative indications, should be $P_f = 10^{-3}$ for limit states which do not endanger lives, and $P_f = 10^{-5} - 10^{-6}$ for disastrous limit states. The corresponding reliabilities would then be 99.9% or 99.9990–99.9999%.
>
> These figures mean that, on average, out of 1,000 nominally identical buildings one will crack or deform excessively, and out of 1,000,000 a number

between 1 and 10 will collapse. Now it is evident that in civil engineering '1,000 identical buildings' (let alone 1,000,000) rarely occur…even neglecting the fact that a statistically significant test would require sample at least 10 or 20 times larger! Moreover, the determination of these low probabilities requires…extrapolations (sometimes drastic) of statistical proprieties that are experimentally known (or can be obtained) only around the *central* values of the random quantities.

For these reasons, the 'probabilities of failure' which can be calculated or estimated in civil engineering have no real *statistical* significance: rather, they conventional, comparative values. Provided this point is clearly understood and accepted, probabilistic method can play a very important role in making rational comparison possible between alternative structural designs. Otherwise, they are vulnerable to all sorts of criticisms."

Grandori (1991) underlines the need for the society to adopt the target reliabilities or tolerated probabilities of failure:

"The probabilistic approach to structural safety is today a well established paradigm. As to the current state of this paradigm , however, one can notice an asymmetry similar to that observed by Freudenthal [see the quote by Freudenthal on p.8], in the traditional approach. An overwhelming part of the research effort, in fact, has been and still is devoted to estimating failure probabilities. By contrast, only sporadic research deals with the problem of choosing an acceptable risk of failure…The concept of structural safety will not leave the 'realm of metaphysics' unless we derive a method for justifying the choice of risk acceptability levels."

The partial answer to this quest was provided by CIRIA-Construction Industry Research and Information Association (1977). They provided with a formula of the tolerable probability of failure P_f of any structure due to any cause in its design life as follows

$$P_f = \frac{10^{-4}\zeta_s T}{L} \tag{2.56}$$

where ζ_s is a social criterion factor, listed in Table 2.1, T is the design life of the structures in years, L is the average number of people within or near the structure during the period of risk.

Table 2.1 Values of the Social Criterion Factor (CIRIA, 1977)

Nature of structure	ζ_s
1. Places of public assembly, dams	0.005
2. Domestic, office, trade and industry	0.05
3. Bridges	0.5
4. Towers, masts, offshore structures	5

In this respect, Augusti, Baratta and Casciati (1984) stress:

"Nevertheless, attempts to make use of present probabilistic approaches, founded on *theoretical* probability of failure P_f of a structure are concerned with the random nature of loads and resistance. But they generally neglect some sources of failure, such as error or negligence, that can be introduced during the design, erection and use of the construction. Therefore, the actual probability of failure P_f' is much larger than the theoretical probability of failure P_f."

CIRIA suggests using the following formula

$$P_f' = 10 P_f \qquad (2.57)$$

Thus the tolerated value of the theoretical probability of failure becomes:

$$P_f = \frac{10^{-5} \zeta_s T}{L} \qquad (2.58)$$

Therefore, if the designed life T is fixed at 50 years and the value $\zeta_s = 0.05$ is taken as the social criterion factor, then

$$P_f = \frac{10^{-5} \times 0.05 \times 50}{L} \approx 10^{-5} - 10^{-7} \qquad (2.59)$$

In Chapter 6 we will illustrate the high sensitivity exhibited by the failure probability.

Neal, Matthews and Vangel (1999) write:

"Reliability methods have been considered for many structural applications including: civil engineering (Cornell, 1969), nuclear reactors (US Nuclear Regulatory Commission, 1975), fixed wing aircraft (Lundberg, 1955), rotorcraft (Arden and Immen, 1988), and space vehicle propulsion systems (Shiao and Chamis, 1989). A reliability goal of 0.999 999 999 jet flight hour was suggested in 1955 by Lundbert for fixed wing civil aircraft... Lincoln (1985) and Corno and Lincoln (1989), using reasoning similar to that of Lundberg (1955) cited reliability goal of 0.9 999 999 per flight for fixed wing military aircraft. The U.S. Army has instituted new structural fatigue integrity criteria for rotorcraft which has been interpreted (Arden and Immen, 1998) as a requirement for a lifetime reliabililty of 0.999 999."

Tichý (1991) stresses;

"Among code makers there exists a natural psychological reluctance to give definite values of P_f and to accept the idea that a certain proportion of constructed facilities will fail. This reluctance to fix P_f strengthens with the growing damage potential of a failure. For these reasons, any recommended values of P_f must be viewed with utmost caution and always in the context of the set of factors affecting the reliability."

For additional pertinent aspects of the acceptable risk, the reader should consult studies by Vrijling (1989), Vrijling, van Hengel and Houben (1998), Brown (1985) and others.

2.6   A Priority Question

The question arises: Who was the first investigator who introduced the concept of safety factor? The account on this topic appears to be inconclusive. Randall (1973, 1976) published two papers on the history of safety factors. The paper (1976) is a condensed version of the earlier one (1973). He mentions (1973, p. 673):

> "In ancient times the safety factor would have been strictly a matter of judgment, based on occasional collapses. Now it is more closely a direct relationship between the strength of materials and stress analysis. Hence, a transition from crude tests, to theories and precise knowledge of materials. This change from margin of ignorance to a calculable factor of safety, marks the beginning of structural safety as we know it today. In homogeneous flexural members it may be said that Coulomb's analysis in 1773 set the stage for formulating modern factors of safety."

Bernshtein (1957) in his book on history of structural mechanics ascribes the concept of safety factors to Navier. Bernshtein (1957 pp. 47-48) discusses in detail the reform in mechanics of solids, introduced by Navier:

> "The essence of this reform, that was understood much later, consisted, firstly, in abandoning the analysis based on the limit state, that was reigning in science from the times of Galilei, and [secondly] to the transfer of analysis based on working state...
>
> The principle of the limit or terminal state is based on the scheme of probable collapse of the structure and determines the value of the load, at which such a collapse can take place. Allowable load is determined by dividing the limit load by the factor of safety.
>
> The principle of working or initial state determines the stress and strain state of the structure during the working, real load, accepting the ideas, that the limit state is fully similar to the working one, such that the ratio between loadings, stresses and displacements on both states is the same and equals to the safety factor.
>
> With such an approach it is not needed to study the limit state; it is sufficient to study the working state, i.e. the stresses and the displacement at the working load, and find their ratio to the limit ones. Since the limit value of the stress for a given material is assumed to be known from experience, which one is being divided by the safety factor produces the so called working stress, the entire analysis reduces to the comparison of real working stresses with the allowable stresses. Therefore, the analysis based on working state is often called the design via the allowable stresses. It must be underlined, that the very concept of the stress was introduced by Navier himself."

Stüssi (1940) (see Straub, 1952, p. 156) wrote:

"The task which Navier set himself, is nothing less than the formulation of a proper method of structural analysis…The fact that we are able today to construct safely and economically, is mainly due to the methods of structural analysis, that particular branch of mechanics which is based on actual working conditions of a structure. These methods were created, within little more than a decade, by a single man, Navier."

Still, some clarifying remarks are called for. Lind writes (2003a):

"the NOTION of the safety factor must be much older than Coulomb. The first time some structures collapsed, somebody must have thought: "If only it had been a little stronger!" The last word implies that strength is a quantity. I also believe Galileo Galilei made strength tests of beams; the intention must have been to establish a load-carrying capacity. To use such data, the safety factor idea pops right out."

Indeed, Galilei was the first who looked for the collapse load, and arrived at a section modulus for the rectangular cross-section to be equal $bh^2/2$, where b is the width and h is the depth. It is of interest that Todhunter and Pierson (1886), Timoshenko (1953) and Szabo (1984) in their definitive accounts on history of mechanics and its principles, do not explicitly speak about the history of the factor of safety. Therefore, it appears that an additional research is needed to establish who first introduced the concept of "safety factor." Possibly the French books on the history of mechanics are more informative in establishing the priority in this field, since both Coulomb and Navier were French.

Mariotte improved Galilei's formula suggesting it to be replaced by $bh^2/3$. Coulomb in 1773 arrived at the correct value of $bh^2/6$. Apparently independently, Navier many years later arrived at the same correct value. Lind (2003b) comments on this issue:
"Anyway, Galilei's erroneous belief that the section modulus is 200% over its true value has no influence on his results; it still gives the correct relationship between width, depth and carrying capacity! A nice example that the wrong theory can give the right results. Another example, one of my favorites, from Martin Gardner's book is the [person] who thought there are four kinds of force: Zig, zag, suction and swirl. He designed aircraft that flew fine."

The book by Heyman (1972) that is dedicated to the Coulomb's memoir on statics, includes both the original French text as well as the English translation. On p. 20 in the French (and on p. 5 4 in English) Coulomb refers to the safety factor concept although not explicitly. After relating the thickness c with the depth b as $c=b/10$, he writes: "If it is desired to increase the mass of the masonry by a quarter above which would be necessary for equilibrium, then $c=b/7$." Thus, the factor of safety is introduced, without direct mentioning of it!

It is quite possible, still, that the notion of safety factor predates both Coulomb (1736-1806) and Navier (1785-1836). Blockley (1980, p.98) writes, referencing Straub (1952, p. 114), about three mathematicians, T. LeSeur, F. Jacquier and R. G. Bosovich, who were asked to examine a particular structure and "to find out the cause of the cracks and

damage which was apparent. This they did in 1742-3 and assessed the value of the the force…by postulating a collapse mechanism and using the equation of virtual work and a safety factor of 2."

Moreover, one of the references in the memoir of Coulomb (see Heyman 1972) is the book by Bélidor (1729). According to Heyman (1972, p. 85), "Bernard Forest de Bélidor (1697-1761) wrote several engineering texts among them the Octavo Dictionnaire portatif de l'ingénieur (1755), and the Architecture Hydraulique (1737-53) in four volumes. The 1729 Science des ingenieurs referred to by Coulomb is also in the nature of an engineer's handbook… Bélidor gives some attention to the factor of safety, and, in a numerical example, he increases the dimensions of the retaining wall to give a safety factor of 1.2." Thus, Bélidor's work appears to be the earliest reference that mentions, according to our research, the notion of the safety factor.

To conclude, it must be stressed that the task of establishing the exact date when the notion of the safety factor was introduced to the engineering practice is escaping us. Although a distinguished colleague of the present writer maintains that the person who introduced the safety factor "was just a cautious man", it is strongly felt that its author was a genius.

2.7 Concluding Comments on the Stress-Strength Interference Method

Although we considered here the stress-strength interface method, the applications of the results reported in this chapter are much broader. As Sundararajan (1995) notes,

"the term "stress" should be considered in a broader sense as any applied load or load-induced response quantity that has the potential to cause failure. Examples are stress, force, moment, torque, pressure, temperature, shock, vibration, stress intensity, strain and deformation. The term "strength" should be considered in a broader sense as the capacity of the component or system to withstand the applied load ("stress"). Examples are yield stress, ultimate stress, yield moment, collapse moment, buckling load, and permissible deformation, depending on the type of applied load (stress, force, moment, deformation, etc.) and the failure criterion (yield failure, collapse, fatigue, excessive deformation, etc.). Some authors use the term "load-capacity interference method", instead of "stress-strength interference method", to indicate the broader scope of the method."

Moreover,

"The stress-strength interference method is one of the earliest methods of structural reliability analysis. Although more advanced and less restrictive methods of reliability analysis have been developed in recent years, the stress-strength interference method is still widely used in many industries because of its simplicity and ease of use."

In their recent definitive monograph Kotz, Lumelskii and Pensky (2003) mention:

"The term "stress" has acquired in the second half of the 20th century a special meaning to a modern person. We all are continuously under stress, and, alas, not always have the "strength" to overcome it. The stress-strength relationship is nowadays studied in many branches of science such as psychology, medicine, pedagogy, etc. and the pharmaceutical industry accumulates one billion-dollar profit assisting not to overcome or at last alleviate psychological stresses.

Broadly speaking, the term stress is used nowadays in two different meanings: 1) structural, mechanical (or engineering) stress studied in the engineering discipline called "strength of materials", and more and more recent concept of 2) psychological stress defined as any "external stimulus - from threatening words to the soul of gunshot – which the brain interprets as dangerous."

Another way of describing this concept is a "demand, threat or other event that requires an individual to cape with a charged situation."

They also write (2003, p.2)

"...the specific practical problem of applied statistics encapsulated by the term "stress-strength"... in the simplest terms...can be described as an assessment of "reliability" of a "component in terms of random variables X representing "stress" experienced by the component and Y representing the "strength" of the component available to overcome the stress. According to this simplified scenario if the stress exceeds the strength ($X>Y$) the component would fail; and vise versa. Reliability is then defined as probability of not failing: $P(X>Y)$."

Kotz, Lumelskii and Pensky (2003) ascribe "the germ of this idea" to Birnbaum (1956) and Birnbaum and McCarthy (1958). They note:

"The formal term "stress-strength" appears in the title of Church and Harris (1970). This is the earliest date in our bibliography though earlier case may exist."

As we see mathematics researchers are unaware of earlier papers by Freudenthal (1947) and Rzhanitsyn (1947) and many others in the mechanics of solids literature! Hopefully this monograph and the bibliography herein will make an additional impact on mathematical investigations in this field.

Prof. Gábor Kazinczy (born in Szeged, Hungary, on January 19, 1889-died in Sweden, on May 26, 1964)

To close this section it must be stated that although Max Mayer was the first one who published a book on probabilistic methods in 1926, Gábor Kazinczy was apparently the first who advocated use of the methods of probability in 1913, although he published his paper only in 1929; he returned to this topic only once, in 1952 (refer to his brief biography in Appendix B).

Chapter 3

Safety Factors and Reliability: Random Actual Stress & Deterministic Yield Stress

"The loads that are acting and/or will act on a structure are never known exactly. They can be estimated from records of loads on similar constructions, but two such records will never be identical. The uncertainty can be reduced but never eliminated..."

G. Augusti, A. Baratta and F. Casciati (1984)

"The factor of safety was a useful invention of the engineer a long time ago that served him well. But it now quite outlived its usefulness and has become a serious threat to real progress in design."

D. Faulkner

"Current structural safety design practices are considered inadequate for future launch vehicles and spacecraft."

V. Verderaime

"[Deterministic safety measures] ignore much information which may be available about uncertainties in structural strengths or applied loads."

R. Melchers

In this chapter we will deal with the simplest possible case, when the yield stress constitutes a deterministic quantity but the actual stress is random, due to the randomness of the distribution or concentrated loads, or the randomness of distributed or the concentrated moments. It becomes possible to express the safety factor via the structural reliability.

3.1   Introductory Comments

Attempts at probabilistic interpretation of the deterministic safety factor have been made in the literature despite the fact that the "spirits" of these two approaches are entirely different. Before discussing them, it is instructive to quote some representative excerpts from popular textbooks concerning its definition, in addition to those given in Section 2.4:

(a)   "To allow for accidental overloading of the structure, as well as for possible inaccuracies in the construction and possible unknown variables in the analysis of the structure, a factor of safety is normally provided by choosing an allowable stress (or working stress) below the proportional limit."

(b)   "Although not commonly used, perhaps a better term for this ratio is factor of ignorance."

(c)   "The need for the safety margin is apparent for many reasons: stress itself is seldom uniform; materials lack the homogeneous properties theoretically assigned to the abnormal loads might occur; manufacturing processes often impart dangerous stresses within the component. These and other factors make it necessary to select working stresses substantially below those known to cause failure."

(d) "A factor of safety is used in the design of structures to allow for (1) uncertainty of loading. (2) the statistical variation of material strengths, (3) inaccuracies in geometry and theory, and (4) the grave consequences of failure of some structures."

(e) "Factor of safety (N_{FS}), where $N_{FS} > 1$, is the ratio of material strength (usually ultimate strength or yield point) to actual or calculated stress. Alternatively, factor of safety can be defined as the ratio of load at failure to actual or calculated load. The factor of safety provides a margin of safety to account for uncetainties such as errors in predicting loading of a part, variations in material properties, and differences between the ideal model and actual material behavior" (Wilson, 1997).

(f) "Choosing the safety factor is often a confusing proposition for the beginning designer. The safety factor can be thought of as a measure of the designer's uncertainty in the analytical models, failure theories, and material property data used, and should be chosen accordingly … Nothing is absolute in engineering any more than in any other endeavor. The strength of materials may vary from sample to sample. The actual size of different examples of the "same" parts made in quantity may vary due to manufacturing tolerance. As a result, we should take the statistical distributions into account in our calculations" (Norton, 2000).

(g) "Most current….vehicles use the Safety Factor approach in the design of structural component. This factor is designed to arbitrarily account for items such as material property variations; manufacturing differences, since no two parts can be made exactly the same; uncertainties in the loading environment; and unknowns in the internal load and stress distributions" (Bruhn, 1975).

(h) "The factor of safety in an arbitrary factor by which the applied loads or load factor are multiplied for the purpose of ensuring sufficient strength to permit the applied loads to be exceeded by a definite amount before complete failure of the structure occurs"(Howell,1956).

3.2   Four Different Probabilistic Definitions of a Safety Factor

The last two quotes in the previous section underline the basic property of the conventional approach to the safety factors, namely, their *arbitrariness*. This section is dedicated to probabilistic reasoning of the safety factors. We start with a simplest possible case of the stress analysis. This is, of course, a member in tension considered by Galileo Galilei some centuries ago (Fig. 3.1). We are interested in the probabilistic generalization of this simplest example. Let an element be subjected to a stress σ. Let it be a random variable, denoted by capital letter Σ, whereas a lower case notation describes the possible values σ that the random variable Σ may take.

Galileo Galilei
(born in Pisa on February 15, 1564, died on Arcetri near Pisa on January 8, 1642)

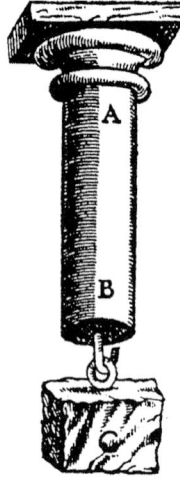

Fig. 3.1: Fundamental problem of probabilistic mechanics is represented by the generalized Galilei problem: a bar under random load

The strength characteristics say the yield stress may also be designated by upper case notation Σ_y, with σ_y being the possible values Σ_y may take. The various possible definitions of the safety factor n are

$$n_1 = \frac{E(\Sigma_y)}{E(\Sigma)} \tag{3.1}$$

which is referred to as the central safety factor; $E(\Sigma_y)$ denotes the mathematical expectation of Σ_y, while $E(\Sigma)$ is associated with the mathematical expectation of Σ.

On the other hand one can treat the ratio

$$Q = \frac{\Sigma_y}{\Sigma} \tag{3.2}$$

as a random variable. Its mathematical expectation

$$n_2 = E(Q) = E\left(\frac{\Sigma_y}{\Sigma}\right) \tag{3.3}$$

could be also interpreted as a safety factor. The third possible definition of the safety factor is

$$n_3 = E(\Sigma_y)E\left(\frac{1}{\Sigma}\right) \tag{3.4}$$

In specific cases some of the above safety factors may coincide. For example, when the yield stress is random, but the stress is a deterministic quantity, *i.e.* takes a single value σ with unity probability, we have

$$n_1 = n_2 = n_3 \tag{3.5}$$

If stress is random, but yield stress is a deterministic quantity σ_y, then

$$n_2 = n_3 \tag{3.6}$$

The fourth definition of the safety factor was proposed by Birger (1970). He considered the probability distribution function $F_Q(q)$ of the random variable $Q = \Sigma_y / \Sigma$

$$F_Q(q) = Prob\left(\frac{\Sigma_y}{\Sigma} \leq q\right) \tag{3.7}$$

Then he demands this function to equal some value p_0:

$$F_Q(q) = Prob\left(\frac{\Sigma_y}{\Sigma} \leq q\right) = p_0 \tag{3.8}$$

The value of $q = q_0$ that corresponds to the $p_0{}^{th}$ fractile of the distribution function $F_Q(q)$ is declared as the safety factor. This implies, that if $p_0 = 0.01$, and say $q_0 = 1.3$, that in about 99% of the realizations of the structure the deterministic safety factor will be not less than 1.3. This factor of safety was independently introduced by Maymon (2002) and will be discussed in some detail in Chapter 7. It must be noted that Bolotin (1975, p.56) in the footnote describes s as a conditional safety factor, whereas the "true" safety factor is a random variable, as he notes.

We are asking ourselves the following question: Can we express the safety factors by probabilistic characterization of the structural performance? The central idea of the probabilistic design of structures is reliability, *i.e.* probability that the structure will perform its mission adequately, as required. In our context the mission itself is defined deterministically, namely we are interested in the event

$$\Sigma < \Sigma_y \tag{3.9}$$

i.e. that the actual stress is less than the yield stress.

Since both Σ and Σ_y may take values from a finite or infinite range of values, the inequality (3.9) will not always take place. For some realizations of random variables Σ and Σ_y the inequality may be satisfied, whereas for the other ones it may be violated. Engineers are interested in the probability that the inequality (3.9) will hold. Such a probability is called *reliability*, denoted by R

$$R = Prob(\Sigma \leq \Sigma_y) \tag{3.10}$$

Its complement

$$P_f = 1 - R = Prob(\Sigma \geq \Sigma_y), \tag{3.11}$$

is called the probability of failure. It is a probability that the stress will be equal to or will exceed the yield stress. It is understandable that engineers want to achieve a very high reliability, allowing, if at all, an extremely small probability of failure.

It appears, at the first glance that the approaches, based on the deterministic allocation of the safety factors, or that based on reliability design are totally contradictory. We will pursue this subject in more detail. In this chapter and in Chapter 4, we discuss particular cases, whereas at a later stage in Chapter 5 we will pursue the general case, in which *both* Σ and Σ_y will be treated as random variables.

3.3   Case 1: Stress Has an Uniform Probability Density, Strength Is Deterministic

Let the stress Σ be a random variable with the uniform probability density

$$f_\Sigma(\sigma) = \begin{cases} \dfrac{1}{\sigma_U - \sigma_L}, & for \ \ \sigma_L < \sigma < \sigma_U \\ 0, & otherwise \end{cases} \tag{3.12}$$

where σ_L is the lowest value that the stress may take, whereas σ_U is the greatest value the stress may assume. We treat the yield stress Σ_y to be a deterministic quantity, *i.e.* to take a single value σ_y with unity probability. The probability distribution function

$$F_\Sigma(\sigma) = Prob(\Sigma \le \sigma) = \int_{-\infty}^{\sigma} f_\Sigma(t) dt \tag{3.13}$$

reads

$$F_\Sigma(\sigma) = \begin{cases} 0, & for \ \ \sigma < \sigma_L \\ \dfrac{\sigma - \sigma_L}{\sigma_U - \sigma_L}, & for \ \ \sigma_L \le \sigma < \sigma_U \\ 1, & for \ \ \sigma_U \le \sigma \end{cases} \tag{3.14}$$

The reliability reads

$$R = Prob(\Sigma \le \sigma_y) \tag{3.15}$$

or, in light of Eq. (3.13) we get

$$R = F_\Sigma(\sigma_y) \tag{3.16}$$

In other words, the reliability equals the stress distribution function F_Σ evaluated at the yield stress (Fig. 3.2).

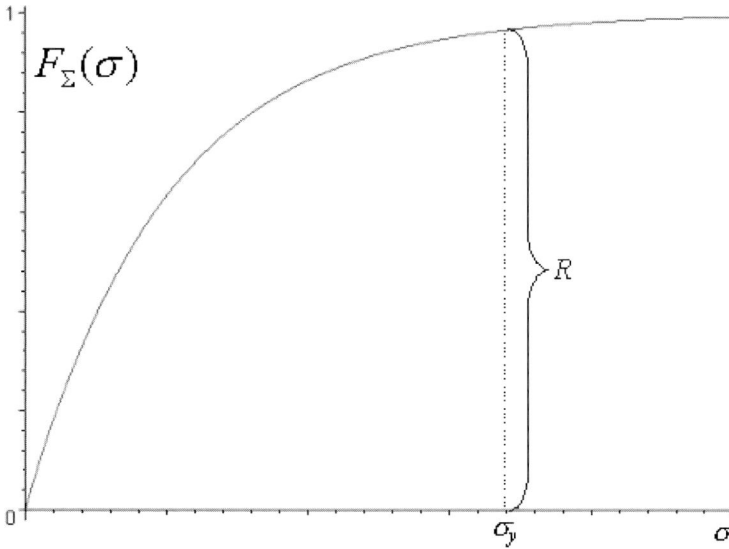

Fig. 3.2 Reliability equals the probability distribution function of the actual stress evaluated at the level of yield stress

Bearing in mind Eq. (3.14) we get

$$R = \begin{cases} 0, & for \ \sigma_y < \sigma_L \\ \dfrac{\sigma_y - \sigma_L}{\sigma_U - \sigma_L}, & for \ \sigma_L \leq \sigma_y < \sigma_U \\ 1, & for \ \sigma_U \leq \sigma_y \end{cases}$$

(3.17)

This formula can be rewritten in more convenient form. We note that the mean value of the stress equals

$$E(\Sigma) = \tfrac{1}{2}(\sigma_L + \sigma_U)$$

(3.18)

whereas the variance of the stress equals

$$Var(\Sigma) = \tfrac{1}{12}(\sigma_U - \sigma_L)^2$$

(3.19)

From these two equations we first express the denominator in Eq. (3.17)

$$\sigma_U - \sigma_L = \sqrt{12 Var(\Sigma)}$$

(3.20)

as well as the lowest possible value the stress can take on

$$\sigma_L = \frac{2E(\Sigma) - \sqrt{12 Var(\Sigma)}}{2}$$

$$= E(\Sigma) - \sqrt{3 Var(\Sigma)} = E(\Sigma) - \sqrt{3}\sigma_\Sigma$$

(3.21)

Hence

$$\sigma_U = E(\Sigma) + \sqrt{3}\sigma_\Sigma \tag{3.22}$$

Let

$$\sigma_L \le \sigma_y < \sigma_U \tag{3.23}$$

Then, in accordance with Eq. (3.17), we have

$$R = \frac{\sigma_y - E(\Sigma) + \sqrt{3Var(\Sigma)}}{2\sqrt{3Var(\Sigma)}} \tag{3.24}$$

By dividing both the numerator and the denominator by the mean value of the actual stress $E(\Sigma)$, and introducing the coefficient of variation of the actual stress $v_\Sigma = \sqrt{Var(\Sigma)}/E(\Sigma)$ we get, instead of Eq. (3.24)

$$R = \frac{n_1 - 1 + \sqrt{3}v_\Sigma}{2\sqrt{3}v_\Sigma} \tag{3.25}$$

As is seen reliability is *directly* expressed in terms of the central safety factor s_1 and the coefficient of variation of the involved random variable v_Σ. Thus, the reliability methods allow to *rigorously*, rather than *arbitrarily* introducing safety factors. The safety factor n_1 corresponding to the required reliability r is expressed as follows

$$n_1 = 1 + \sqrt{3}v_\Sigma (2r - 1) \tag{3.26}$$

Maximum value of the safety factor $n_{1,max} = 1 + \sqrt{3}v_\Sigma$ is achieved when the reliability tends to unity from below. For example, for coefficient of variation 0.05 the safety factor assumes the value 1.09; for the coefficient of variation 0.1 the safety factor equals 1.17; for coefficient of variation 0.15 it takes a value 1.26 *etc.* We conclude that with greater variation of the involved random variable, the safety factor must be *increased*. This *qualitative* conclusion is in line with our anticipation. Yet, it is seen that the reliability context allows one to make *quantitative* judgments in terms of the *required reliability* and the *coefficient of variability.*

Note that this formula is valid when the inequality (3.23) holds. This inequality is rewritten as follows, in view of Eq (3.21) and Eq (3.22):

$$1 - \sqrt{3}\ v_\Sigma \le n_1 = 1 - \sqrt{3}\ v_\Sigma + 2\sqrt{3}\ v_\Sigma r \le 1 + \sqrt{3}\ v_\Sigma \tag{3.27}$$

Let us check if the expressions (3.27) and (3.28) are compatible with each other. Since $2r-1$ does not exceed unity, the right hand side of the inequality (3.28) is compatible with Eq. (3.27). So is the left hand side: The first two terms coincide with each other, on the both sides of inequality. We conclude that the expression (3.27) is compatible with the restriction in Eq. (3.23) as it should be.

3.4 Case 2: Stress Has an Exponential Probability Density, Yield Stress Is Deterministic

Consider now that the stress has an exponential probability density

$$f_\Sigma(\sigma) = \begin{cases} 0, & for \ \sigma < 0 \\ a\exp(-a\sigma), & for \ \sigma \geq 0 \end{cases} \tag{3.28}$$

The corresponding probability distribution function reads

$$F_\Sigma(\sigma) = Prob(\Sigma \leq \sigma_y) = [1 - \exp(-a\sigma)]U(\sigma) \tag{3.29}$$

where $U(\sigma)$ is the unit step function; it equals unity for positive γ and vanishes otherwise. The parameter a is reciprocal to mean value of stress

$$E(\Sigma) = \frac{1}{a} \tag{3.30}$$

Also, since parameter a is the only free parameter in the density (3.28) *all* probabilistic moments depend solely upon it. Thus, variance also is expressible in terms of a, as follows:

$$Var(\Sigma) = \frac{1}{a^2} \tag{3.31}$$

Since the coefficient of variation

$$v_\Sigma = \frac{\sqrt{Var(\Sigma)}}{E(\Sigma)} = \frac{1/a}{1/a} = 1 \tag{3.32}$$

is unity, or 100%, we must anticipate high levels of safety factor, in order to ensure the high level of required reliability. The latter equals, in view of Eq. (3.19)

$$\begin{aligned} R &= Prob(\Sigma \leq \sigma_y) \\ &= [1 - \exp(-a\sigma_y)]U(\sigma_y) \end{aligned} \tag{3.33}$$

We first express the value of a from Eq. (3.30) as

$$a = \frac{1}{E(\Sigma)} \tag{3.34}$$

and substitute it into Eq. (3.33), to arrive at

$$R = \left[1 - \exp\left(-\frac{\sigma_y}{E(\Sigma)}\right)\right]U(\sigma_y) \tag{3.35}$$

In view of the central safety factor s_1, Eq. (3.35) is rewritten as

$$R = [1 - \exp(-s)]U(\sigma_y) \tag{3.36}$$

As is seen, a *direct* relationship is being established between the safety factor and the reliability. Once the required reliability is specified the associated safety factor equals

$$n = \ln \frac{1}{1 - R} \tag{3.37}$$

For example, reliability of 0.9 leads to the safety factor 2.3; the reliability of 0.95 results in safety 3 *etc.* Such high values, as indicated above stem from the fact that the stress exponential probability density is associated with high, namely 100% variability. It is immediately seen, that one of the reasons for the high variations in this particular case is the fact that the stress can take any value on the positive axis.

3.5 Case 3: Stress Has a Rayleigh Probability Density, Yield Stress is Deterministic

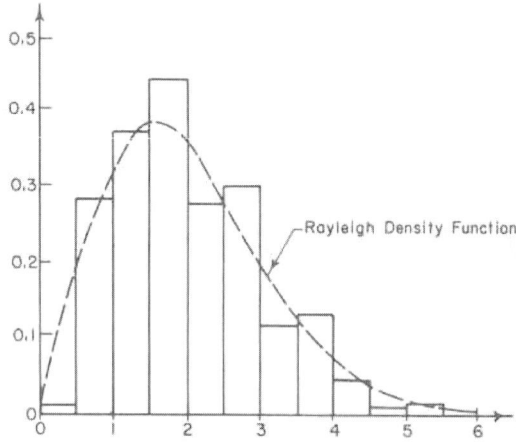

Fig. 3.3 Rayleigh density function

Consider now the case in which the stress is characterized by a Rayleigh probability density:

$$f_\Sigma(\sigma) = \begin{cases} 0, & for \ \sigma < 0 \\ \dfrac{\sigma}{b^2} \exp(-\dfrac{\sigma^2}{2b^2}), & for \ \sigma \geq 0 \end{cases} \tag{3.38}$$

The approximate probability distribution function is

$$F_\Sigma(\sigma) = Prob(\Sigma \leq \sigma) = \int_{-\infty}^{\sigma} f_\Sigma(\alpha)d\alpha = \left[1 - \exp\left(-\frac{\sigma^2}{2b^2} \right) \right] U(\sigma) \tag{3.39}$$

The reliability evaluation reads:

$$R = Pr\,ob(\Sigma \leq \sigma_y) = F_\Sigma(\sigma_y) \tag{3.40}$$

Again, the reliability equals the probability distribution function of the *stress evaluated at the level of the yield stress*. Hence, in view of Eq. (3.39)

$$R = \left[1 - \exp\left(-\frac{\sigma_y^{\,2}}{2b^2}\right)\right] U(\sigma_y) \tag{3.41}$$

We would like now to express parameter b in Eqs. (3.38) and (3.39) through the probabilistic characterization of the stress:

$$E(\Sigma) = \int_{-\infty}^{\infty} \sigma f_\Sigma(\sigma) d\sigma = \frac{b\sqrt{\pi}}{\sqrt{2}} \approx 1.25b \tag{3.42}$$

$$Var(\Sigma) = \int_{-\infty}^{\infty} (\sigma - E(\Sigma))^2 f_\Sigma(\sigma) d\sigma$$

$$= \frac{4-\pi}{2} b^2 \approx 0.43b^2 \tag{3.43}$$

We express b from Eq. (3.42)

$$b \approx \frac{E(\Sigma)}{1.25} = 0.8 E(\Sigma) \tag{3.44}$$

and substitute it into Eq. (3.41) to yield

$$R = \left\{1 - \exp\left(-\frac{\sigma_y^{\,2}}{2[0.8E(\Sigma)]^2}\right)\right\} U(\sigma_y)$$

$$= \left\{1 - \exp\left(-\frac{0.78125\sigma_y^{\,2}}{[E(\Sigma)]^2}\right)\right\} U(\sigma_y) \tag{3.45}$$

We take into account the definition of the central safety factor n_1 to get

$$R = [1 - \exp(-0.78125 n_1^{\,2})] U(\sigma_y) \tag{3.46}$$

This formula allows expressing the safety factor by the reliability

$$n \approx 1.13 \sqrt{\ln\frac{1}{1-R}} \tag{3.47}$$

Thus, the reliability $R = 0.9$ yields in central safety factor 1.71, the reliability of 0.95 results in safety factor 1.96. The required reliability of 0.99 is associated with safety factor 2.42 *etc.* Again, reason for these values is the high coefficient of variation. Indeed, Eqs. (3.42) and (3.43) suggest that the coefficient of variation equals:

$$v_\Sigma = \frac{\sqrt{Var(\Sigma)}}{E(\Sigma)} \approx \frac{\sqrt{0.43b^2}}{1.25b} = 0.52 \tag{3.48}$$

Although this is a smaller variability than in the case of the stress with exponential probability density, still, hopefully, 52% variation is seldom encountered in practice.

3.6   Case 4: Stress Has a Normal Probability Density, Yield Stress Is Deterministic

We consider now the case in which the stress is characterized by a normal probability density

$$f_\Sigma(\sigma) = \frac{1}{b\sqrt{2\pi}} \exp\left[-\frac{1}{2}\left(\frac{\sigma-a}{b}\right)^2\right], \quad -\infty < \sigma < \infty \tag{3.49}$$

The distribution function reads

$$F_\Sigma(\sigma) = \frac{1}{b\sqrt{2\pi}} \int_{-\infty}^{\sigma} \exp\left[-\frac{1}{2}\left(\frac{t-a}{b}\right)^2\right] dt$$

$$= \Phi\left(\frac{\sigma-a}{b}\right) \tag{3.50a}$$

$$\Phi(x) = \frac{1}{\sqrt{2\pi}} \int_{-\infty}^{x} \exp\left(-\frac{1}{2}t^2\right) dt \tag{3.50b}$$

where a is the mean stress, and b is the mean square deviation,

$$E(\Sigma) = a \tag{3.51}$$

$$Var(\Sigma) = b^2 \tag{3.52}$$

The reliability equals

$$R = Prob(\Sigma < \sigma_y) = F_\Sigma(\sigma_y) = \Phi\left(\frac{\sigma_y - a}{b}\right) \tag{3.53}$$

or with Eq. (3.50) taken into account

$$R = \Phi\left(\frac{\sigma_y - E(\Sigma)}{\sqrt{Var(\Sigma)}}\right) \tag{3.54}$$

Dividing both the numerator and the denominator by $E(\Sigma)$ we rewrite Eq. (3.53) as follows:

$$R = \Phi\left(\frac{n_1 - 1}{v_\Sigma}\right) \tag{3.55}$$

where n_1 is the central safety factor, v_Σ is the coefficient variation of the stress. Eq. (3.54) allows again to express the safety factor via the reliability

$$n_1 = 1 + v_\Sigma \Phi^{-1}(R) \tag{3.56}$$

where $\Phi^{-1}(R)$ is a function that is inverse to $\Phi(R)$.

We note the following values of the inverse normal probability function (Benjamin and Cornell, 1970, p. 655)

$$\Phi^{-1}(0.9) = \Phi^{-1}(1-10^{-1}) = 1.28$$

$$\Phi^{-1}(0.99) = \Phi^{-1}(1-10^{-2}) = 2.32$$

$$\Phi^{-1}(0.999) = \Phi^{-1}(1-10^{-3}) = 3.09$$

$$\Phi^{-1}(0.9\ 999) = \Phi^{-1}(1-10^{-4}) = 3.72$$

$$\Phi^{-1}(0.99\ 999) = \Phi^{-1}(1-10^{-5}) = 4.27$$

$$\Phi^{-1}(0.999\ 999) = \Phi^{-1}(1-10^{-6}) = 4.75 \qquad (3.57)$$

$$\Phi^{-1}(0.9\ 999\ 999) = \Phi^{-1}(1-10^{-7}) = 5.20$$

$$\Phi^{-1}(0.99\ 999\ 999) = \Phi^{-1}(1-10^{-8}) = 5.61$$

$$\Phi^{-1}(0.999\ 999\ 999) = \Phi^{-1}(1-10^{-9}) = 6.00$$

$$\Phi^{-1}(0.9\ 999\ 999\ 999) = \Phi^{-1}(1-10^{-10}) = 6.36$$

$$\Phi^{-1}(0.99\ 999\ 999\ 999) = \Phi^{-1}(1-10^{-11}) = 6.71$$

Thus, the safety factor becomes, for the coefficient of variation equal 0.05, respectively

$$n_1 = 0.05\Phi^{-1}(0.9) = 1.064, \quad for \quad R = 1-10^{-1}$$

$$n_1 = 0.05\Phi^{-1}(0.99) = 1.116, \quad for \quad R = 1-10^{-2}$$

$$n_1 = 0.05\Phi^{-1}(0.999) = 1.155, \quad for \quad R = 1-10^{-3}$$

$$n_1 = 0.05\Phi^{-1}(0.9\ 999) = 1.186, \quad for \quad R = 1-10^{-4}$$

$$n_1 = 0.05\Phi^{-1}(0.99\ 999) = 1.2135, \quad for \quad R = 1-10^{-5}$$

$$n_1 = 0.05\Phi^{-1}(0.999\ 999) = 1.2375, \quad for \quad R = 1-10^{-6} \qquad (3.58)$$

$$n_1 = 0.05\Phi^{-1}(0.9\ 999\ 999) = 1.26, \quad for \quad R = 1-10^{-7}$$

$$n_1 = 0.05\Phi^{-1}(0.99\ 999\ 999) = 1.2805, \quad for \quad R = 1-10^{-8}$$

$$n_1 = 0.05\Phi^{-1}(0.999\ 999\ 999) = 1.3, \quad for \quad R = 1-10^{-9}$$

$$n_1 = 0.05\Phi^{-1}(0.9\ 999\ 999\ 999) = 1.318, \quad for \quad R = 1-10^{-10}$$

$$n_1 = 0.05\Phi^{-1}(0.99\ 999\ 999\ 999) = 1.3355, \quad for \quad R = 1-10^{-11}$$

As is seen, there is direct relationship between the safety factor and required reliability. One can suggest asymptotic relationship between the safety factor and reliability. One observes from Eq. (3.57) that the knowledge of the coefficient of variation (the ratio between the standard deviation and the mean) and required reliability directly yields the level of the required safety factor.

3.7 Case 5: Actual Stress Has a Log-Normal Probability Density, Yield Stress is Deterministic

Consider now the case in which the actual stress Σ is distributed log-normally, with the following probability density function:

$$f_{\Sigma}(\sigma) = \frac{1}{\sigma b_{\Sigma}\sqrt{2\pi}}\exp\left[-\frac{(\ln\sigma - a_{\Sigma})^2}{2b_{\Sigma}^{2}}\right], \quad \sigma > 0 \tag{3.59}$$

and vanishes otherwise. The mean value of the stress equals

$$E(\Sigma) = \exp(a_{\Sigma} + \tfrac{1}{2}b_{\Sigma}^{2}) \tag{3.60}$$

whereas the variance reads:

$$Var(\Sigma) = \exp(2a_{\Sigma} + b_{\Sigma}^{2})[\exp(b_{\Sigma}^{2}) - 1] \tag{3.61}$$

The reliability equals

$$R = Prob(\Sigma \le \sigma_y) = F_{\Sigma}(\sigma_y) \tag{3.62}$$

The probability distribution function for the log-normal variable Σ is

$$F_{\Sigma}(\sigma) = \int_{0}^{\sigma}\frac{1}{t b_{\Sigma}\sqrt{2\pi}}\exp\left[-\left(\frac{\ln t - a_{\Sigma}}{b_{\Sigma}}\right)^2\right]dt \tag{3.63}$$

We make a substitution

$$\frac{\ln t - a_{\Sigma}}{b_{\Sigma}} = z \tag{3.64}$$

to obtain

$$F_{\Sigma}(\sigma) = \int_{0}^{\frac{\ln\sigma - a_{\Sigma}}{b_{\Sigma}}}\frac{1}{\sqrt{2\pi}}\exp\left(-\tfrac{1}{2}z^2\right)dt$$
$$= \Phi\left(\frac{\ln\sigma - a_{\Sigma}}{b_{\Sigma}}\right) \tag{3.65}$$

In view of Eq. (3.65), expression for the reliability

$$R = \Phi\left(\frac{\ln\sigma_y - a_{\Sigma}}{b_{\Sigma}}\right) \tag{3.66}$$

Central safety factor equals

$$n_1 = \frac{\sigma_y}{E(\Sigma)} = \frac{\sigma_y}{\exp\left(a_{\Sigma} + \dfrac{1}{2}b_{\Sigma}^{2}\right)} \tag{3.67}$$

Knowing parameters a_{Σ} and b_{Σ} determines both the central safety factor and the reliability.

On the other hand, if $E(\Sigma)$ and $Var(\Sigma)$ are given, one needs the formula of transformation from Eqs. (3.60) and (3.61). We substitute Eq. (3.67) into Eq. (3.66) to get

61

$$R = \Phi\left(\frac{\ln\sigma_y - \ln E(\Sigma) - \frac{1}{2}\{\ln[E^2(\Sigma)+Var(\Sigma)]-\ln[E^2(\Sigma)]\}}{\sqrt{\ln[E^2(\Sigma)+Var(\Sigma)]-\ln[E^2(\Sigma)]}}\right) \tag{3.68}$$

This formula can be rewritten in a more concise form:

$$R = \Phi\left(\frac{\ln n_1 - \frac{1}{2}[\ln(1+v_\Sigma^2)]}{\sqrt{\ln(1+v_\Sigma^2)}}\right) \tag{3.69}$$

3.8 Case 6: Actual Stress Has a Weibull Probability Density, Strength is Deterministic

Let us study the case of the probability of the distribution function stress

$$F_\Sigma(\sigma) = \exp\left[-\exp\left(\frac{a_\Sigma-\sigma}{b_\Sigma}\right)\right] \tag{3.70}$$

The reliability, therefore, is given by

$$R = F_\Sigma(\sigma_y) = \exp\left[-\exp\left(\frac{a_\Sigma-\sigma_y}{b_\Sigma}\right)\right] \tag{3.71}$$

According to Haldar and Mahadevan (2000) who do not deal with the material in this section, but use the Weibull distribution, the mean value and the variance can be expressed via a_Σ and b_Σ analytically. In our setting their formulas read:

$$\frac{1}{b_\Sigma} = \frac{1}{\sqrt{\sigma}}\frac{\pi}{\sqrt{Var(\Sigma)}} \tag{3.72}$$

$$a_\Sigma = E(\Sigma) - 0.5772 b_\Sigma$$

Thus,

$$E(\Sigma) = a_\Sigma + 0.5772 b_\Sigma \tag{3.73}$$

$$Var(\Sigma) = \frac{\pi^2}{6}b_\Sigma^2 \tag{3.74}$$

Reliability becomes:

$$R = \exp\left[-\exp\left(\frac{E(\Sigma) - 0.5772\frac{\sqrt{6Var(\Sigma)}}{\pi} - \sigma_y}{\frac{\sqrt{6Var(\Sigma)}}{\pi}}\right)\right] \qquad (3.75)$$

or, dividing both numerator and denominator by $E(\Sigma)$ and recalling definition of the central safety factor $n_1 = \sigma_y / E(\Sigma)$ and of the coefficient of variation of the actual stress $v_\Sigma = \sqrt{Var(\Sigma)} / E(\Sigma)$ we get

$$R = \exp\left[-\exp\left(\frac{1 - 0.45v_\Sigma - n_1}{0.78v_\Sigma}\right)\right] \qquad (3.76)$$

This formula too apparently is given for the first time. It connects the reliability with the central safety factor n_1 and the variability v_Σ. Conversely, if the required reliability is specified, one can directly determine the safety factor

$$n_1 = 1 - v_\Sigma\left[0.78\ln\left(\ln\frac{1}{R}\right) - 0.45\right] \qquad (3.77)$$

The following values are obtained for $v_\Sigma = 0.05$:

$$\begin{array}{ll}
R = 0.9, & n_1 = 1.11 \\
R = 0.95, & n_1 = 1.14 \\
R = 0.99, & n_1 = 1.18 \\
R = 0.999, & n_1 = 1.29 \\
R = 0.9999, & n_1 = 1.36 \\
R = 0.99999, & n_1 = 1.45
\end{array} \qquad (3.78)$$

For $v_\Sigma = 0.1$ we get

$$\begin{array}{ll}
R = 0.9, & n_1 = 1.22 \\
R = 0.95, & n_1 = 1.28 \\
R = 0.99, & n_1 = 1.36 \\
R = 0.999, & n_1 = 1.58 \\
R = 0.9999, & n_1 = 1.72 \\
R = 0.99999, & n_1 = 1.90
\end{array} \qquad (3.79)$$

Note this , the Gumbel distribution variable is also said to have the type I extreme value distribution. The y are also type II and type III extreme value distribution.

3.9 Actual Stress Has a Fréchet Probability Distribution , Yield Stress Is Deterministic

The probability distribution function $F_\Sigma(\sigma)$ of Maurice Fréchet (1878-1973) reads

$$F_\Sigma(\sigma) = e^{(\frac{a}{\sigma})^k} \tag{3.80}$$

here a and k are the parameter of the distribution, k in a measure of dispersion.
Hence the reliability is:

$$R = F_\Sigma(\sigma_y) = \exp\left[-\left(\frac{a}{\sigma_y}\right)^k\right]^\Sigma, \tag{3.81}$$

The mean value $E(\Sigma)$ of stress reads:

$$E(\Sigma) = a\Gamma\left(1 - \frac{1}{k}\right) \tag{3.82}$$

whereas the variance equals,

$$Var(\Sigma) = a^2\left[\Gamma(1 - \frac{2}{k}) - \Gamma^2(1 - \frac{1}{k})\right], \tag{3.83}$$

where Gamma function is defined as an integral,

$$\Gamma(z) = \int_0^\infty t^{z-1}e^{-t}dt \tag{3.84}$$

for integer value of x

$$\Gamma(1+x) = x\Gamma(x) = x! \tag{3.85}$$

For non integer value, one can use the polynomial approximation:

$$\Gamma(1+x) = 1 + b_1 x + b_2 x^2 + b_3 x^3 + b_4 x^4 + b_5 x^5 \tag{3.86}$$

$b_1 = -0.577191652, \ b_2 = 0.988205891, \ b_3 = -0.897056937,$

$b_4 = 0.918206857, \quad b_5 = -0.756704078, \quad b_6 = 0.482199384,$

$b_7 = -0.193527818, \quad b_8 = 0.035868343$

Haldar and Mahadevan (2000) evaluate an example:

64

$$\Gamma(11/8) = \Gamma(1+3/8) = 1 - 0.577191652\,(3/8)$$
$$+ 0.988205891\,(3/8)^2 + ...$$
$$+ 0.035868343\,(3/8)^8 =$$
$$= 0.888913365$$

(3.87)

Expressing 'a' from Σ_y (382)

$$a = \frac{E(\Sigma)}{\Gamma(1-\frac{1}{k})}$$

(3.88)

and substituting into Σ_y (0.8) and obtain

$$R = \exp\left[-\frac{E^k(\Sigma)}{\Gamma^k(1-\frac{1}{k})\sigma_y^{\,k}}\right]$$

(3.99)

Observing that the ratio

$$\frac{\sigma_y}{E(\Sigma)} = s$$

(3.100)

is nothing else but the central safety factor, the reliability R can be cast in the following form

$$R = \exp\left[-\frac{1}{\Gamma^k(1-\frac{1}{k})n^k}\right]$$

(3.101)

i.e. the direct connection is established between the reliability and the central safety factor n.

If the required reliability is r the Eq. (3.101) immediately provides the central safety factor n that ensures the reliability level r:

$$n = \frac{1}{\Gamma\left(1-\frac{1}{k}\right)}\sqrt[k]{-\frac{1}{\ln r}}$$

(3.102)

if $k = \frac{8}{3}, \Gamma\left(1-\frac{3}{8}\right) = \Gamma\left(\frac{5}{8}\right) = 1.428045305$. Calculations yield:

For $r = 0.9$, $n = 1.62837$;

for $r = 0.95$, $n = 2.13297$;

for $r = 0.99$, $n = 3.93043$;

where $\Gamma(x)$, can be estimated by using spreadsheet (Haldar and Mahadevan, 2000):

Microsoft Excel – EXP(GAMMALN(z))

EXP (GAMMALN (11/8)) = 0.888913569,

EXP (GAMMALN (5/8)) = 1.434518848,

EXP (GAMMALN (5.2)) = 32.57809604

Quattro Pro - @EXP (@ GAMMALN(x))

@ EXP (@ GAMMALN (11/8)) = 0.88891336,

@ EXP (@ GAMMALN (5/8)) = 1.43451904,

@ EXP (@ GAMMALN (5.2)) = 32.57809603.

Table 3.1. Reliability vs. safety factor for different values of k (Fréchet distributed stress)

r	k	$1-\dfrac{1}{k}$	$\Gamma\left(1-\dfrac{1}{k}\right)$	n	r	n	r	n	r	n
	1.5	0.33333	2.67894	1.673		2.704		8.015		37.316
	2.0	0.5	1.77245	1.738		2.491		5.628		17.837
0.9	3.0	0.66667	1.35412	1.564	0.95	1.988	0.99	3.422	0.999	7.384
	4.0	0.75	1.22542	1.432		1.715		2.577		4.588
	5.0	0.8	1.16423	1.347		1.556		2.155		3.419

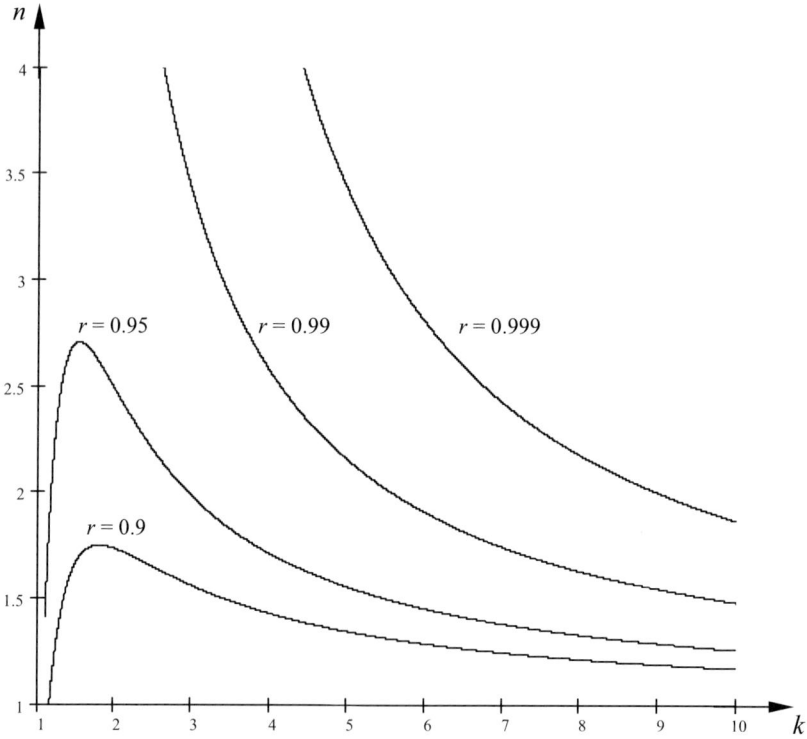

Fig. 3.4 Central safety factor n vs. coefficient k, for various required reliabilities r

3.10   Case 7: Actual Stress Has Two Parameter Weibull Probability density, Yield Stress Is Deterministic

So-called two-parameter Weibull probability distribution for the actual stress is given by;

$$F_{\Sigma}(\sigma) = 1 - \exp\left[-\left(\frac{\sigma}{c_{\Sigma}}\right)^{k}\right], \sigma \geq 0 \qquad (3.103)$$

where k and c_{Σ} are positive constants. The mean value of $E(\Sigma)$ reads:

$$E(\Sigma) = c_{\Sigma}\Gamma\left(1 + \frac{1}{k}\right) \qquad (3.104)$$

whereas the variance is

$$\text{Var}(\Sigma) = c_{\Sigma}\left[\Gamma\left(1 + \frac{2}{k}\right) - \Gamma^{2}\left(1 + \frac{1}{k}\right)\right] \qquad (3.105)$$

The reliability R is obtained by substituting formally $\sigma = \sigma_{y}$ in Equation (3.103).

$$R = 1 - \exp\left[-\left(\frac{\sigma_{y}}{c_{\Sigma}}\right)^{k}\right], \ \sigma_{y} \geq 0 \qquad (3.106)$$

we express c from Equation (3.104) and substitute into Eq. (3.106) to yield:

$$R = 1 - \exp\left\{-\left[\sigma_{y}\frac{\Gamma\left(1 + \frac{1}{k}\right)}{E(\Sigma)}\right]\right\} \qquad (3.107)$$

In view of the definition of the central safety factor $n = \sigma_{y}/E(\Sigma)$ we write:

$$R = 1 - \exp\left\{-\left[n\Gamma\left(1 + \frac{1}{k}\right)\right]\right\} \qquad (3.108)$$

For the required reliability r the needed central safety factors is explicitly found:

$$n = \frac{\ln\left[\frac{1}{(1-r)}\right]}{\Gamma\left(1 + \frac{1}{k}\right)}, \qquad (3.109)$$

Hence, for $r = 0.9$ and $k = 0.5$, $\Gamma(3) = 2$, we get $n = 1.5129$. Furthermore, for $r = 0.95$, $n = 1.62837$;

for $r = 0.99$, $n = 2.13297$;

for $r = 0.999$, $n = 3.930443$.

Note that if $k = 1$, we get the exponentially distributed actual stress, moreover, if $k = 2$ the actual stress has a Rayleigh distribution.

Table 3.2. Reliability vs. safety factor for various values of k (Weibull probability distribution with two parameters).

r	k	$1+\dfrac{1}{k}$	$\Gamma\left(1+\dfrac{1}{k}\right)$	n	r	n	r	n	r	n
	0.5	3	2	1.151		1.498		2.303		3.454
	1.0	2	1	2.303		2.996		4.605		6.908
0.9	2.0	1.5	0.86623	2.598	0.95	3.38	0.99	5.196	0.999	7.795
	3.0	1.33333	0.89298	2.579		3.355		5.157		7.736
	4.0	1.25	0.90640	2.54		3.305		5.081		7.621

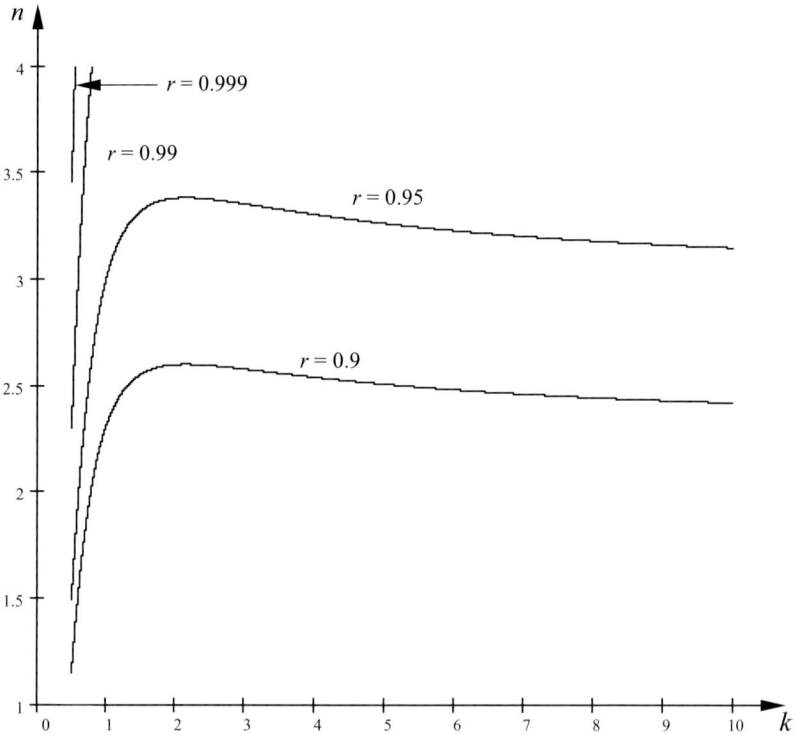

Fig. 3.5 Central safety factor n vs. coefficient k, for various required reliabilities r

3.11   Case 8: Actual Stress Has a Three Parameter Weibull Probability Density, Yield Stress Is Deterministic

The three-parameter Weibull distribution for the actual stress Σ is given by

$$F_{\Sigma}(\sigma) = 1 - \exp\left\{-\left[\left(\frac{\sigma - \sigma_0}{d - \sigma_0}\right)\right]^k\right\}, (\sigma \geq \sigma_0) \tag{3.110}$$

where σ_0 = minimum possible value of $\Sigma \geq 0$, d = characteristic parameter ($d \geq \sigma_0$) and k = shape parameter ($k>0$) indicate the three parameters of the distribution. If the minimum stress σ_0 is fixed at zero, Eq. (3.110) became a two-parameter Gumbel distribution.

The mean value $E(\Sigma)$ of the stress reads:

$$E(\Sigma) = \sigma_0 + (d - \sigma_0)\Gamma\left(1 + \frac{1}{k}\right) \tag{3.111}$$

whereas the variance equals:

$$Var(\Sigma) = (d - \sigma_0)^2\left[\Gamma\left(1 + \frac{2}{k}\right) - \Gamma^2\left(1 + \frac{1}{k}\right)\right], \tag{3.112}$$

The reliability R becames,

$$R = F_{\Sigma}(\sigma_y) = 1 - \exp\left\{-\left[\left(\frac{\sigma_y - \sigma_0}{d - \sigma_0}\right)\right]^k\right\}, (\sigma_y \geq \sigma_0) \tag{3.113}$$

We express the quantity $d - \sigma_0$ from Eq. (3.111) and substitute into Eq. (3.113) to yield:

$$R = 1 - \exp\left(-\left\{(\sigma_y - \sigma_0)\Gamma\frac{\left(1 + \frac{1}{k}\right)}{[E(\Sigma) - \sigma_0]}\right\}^k\right) \tag{3.114}$$

Or, after some algebra, the reliability expression becomes

$$R = 1 - \exp\left\{-\left[(s - \gamma)\Gamma\frac{\left(1 + \frac{1}{k}\right)}{(1 - \gamma)}\right]^k\right\} \tag{3.115}$$

$$\gamma = \frac{\sigma_0}{E(\Sigma)}$$

The safety factor n is explicitly obtainable for the required reliability level $R = r$:

$$n = \gamma + (1 - \gamma)\left(\ln \frac{1}{1-r} \right)^{\frac{1}{k}} \frac{1}{\Gamma\left(1 + \frac{1}{k}\right)}$$

(3.116)

for $k = \frac{1}{3}$, we get:

for $r = 0.9$, $n = 1.01513$;

for $r = 0.95$, $n = 1.04979$;

for $r = 0.99$, $n = 1.13026$;

for $r = 0.999$, $n = 1.24539$.

Table 3.3. **Central safety factor for various values of k (three parameter Weibull distribution $\gamma = 0.9$).**

r	k	$1+\dfrac{1}{k}$	$\Gamma\!\left(1+\dfrac{1}{k}\right)$	n	r	n	r	n	r	n
	0.5	3	2	1.151		1.498		2.303		3.454
	1.0	2	1	2.303		2.996		4.605		6.908
0.9	2.0	1.5	0.86623	2.598	0.95	3.38	0.99	5.196	0.999	7.795
	3.0	1.33333	0.89298	2.579		3.355		5.157		7.736
	4.0	1.25	0.90640	2.54		3.305		5.081		7.621

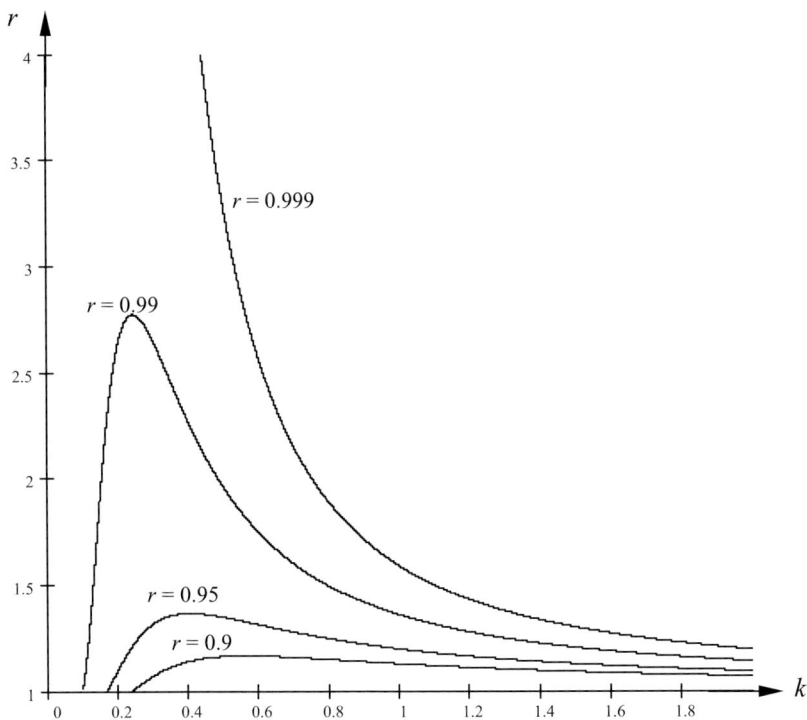

Fig. 3.6 Central safety factor n vs. coefficient k, for various required reliabilities r

3.12   Discussion: Augmenting Classical Safety Factors via Reliability

As is observed from this chapter the use of the safety factor is not contradictory to the employment of the probabilistic methods. Moreover, in many cases the safety factors can be directly expressed by the required reliability levels. However, there is a major difference that must be emphasized: Whereas the safety factors are allocated in an *ad hoc* manner, the probabilistic approach offers a unified mathematical framework. The establishment of the interrelation between the concepts opens an avenue for rational of safety factors, based on reliability.

If there are several forms of failure then the allocation of safety factors should be based on having the *same* reliability associated with each failure modes. This immediately suggests that by the probabilistic methods the existing overdesign or underdesign can be eliminated.

This is done by calibration of the reliability levels with one of the safety factors that is already accepted. Thus, via such an approach, the other failure modes' safety factors can be established.

This is illustrated Fig. 3.3, which shows that presently safety factor are assigned in an *ad hoc* manner to each failure mode, but there is no interrelation between them. Fig. 3.4 illustrates that the consistent allocation of the safety factors can be performed. It is instructive to note that Freudenthal (1956) advocated that actually it was better to use the safety factor terminology even within the realism of probabilistic design:

> "….because the concept of safety is deeply rooted in engineering design whereas the notion that a finite (no matter how small) probability of failure or at least of unserviceability is repulsive to a majority of engineers, it appears desirable to return the concept of "safety" rather than to replace it by that of "probability of failure" and to reformulate the former in terms of the latter."

"Should then the traditional concept of the safety factor remain intact?" the surprised reader may ponder. Freudenthal (1956) responds to such a possible inquiry:

> "The safety factor is thus transformed into a parameter that is a function of the random variation of all design characteristics as well as of the non-random variation essentially caused by the process of construction. Only the part representing the random variations can be derived from the relevant probabilities; that part represents, therefore, the objective minimum value of the safety factor. The selection of the actual engineering safety factor should be guided by this minimum value in conjunction with sound judgment concerning the possible uncertainties associated with actual construction."

It should be noted that the above ideas correlate well with the philosophy developed by Streletskii (1937):

"The factor of safety, i.e. the ratio of ultimate load to the working load, is the fundamental coefficient for structural design. The factor of safety is usually considered to be the coefficient of strength equality of structures… This principle is used in designing working loads in all fields of engineering practice in all countries. Analysis shows, however, that the method in question does not agree with conditions existing in reality. This is due to an insufficiently clear understanding of the term "strength equality"…strength depends on chances of failure and equal strength depends on equal chances, or more correctly, on equal probability of failure. Those structures may be called structures of equal strength which have an equal probability of failure…The factor of safety is thus by no means a coefficient of strength equality."

In his polemic article Tal (1970), quoting from Gemmerling (1974) stated that at present time the role of the probabilistic analysis of structures should consist in eliminating the lack of equal reliability of various structures.

The Chapter 4 will discuss the reverse case, namely when the actual stress is deterministic, but the yield stress is random. Chapter 5 will discuss the general case in which *both* the actual stress and the yield stress are treated as random quantities with the attendant interrelationship between the reliability and safety factors.

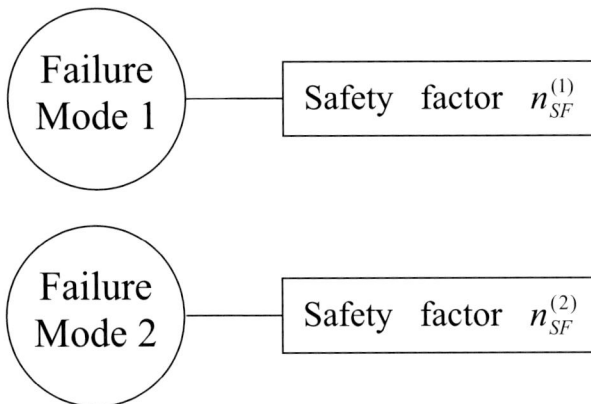

Fig. 3.9 Present status: no connection between safety factors, leading to either overdesign or underdesign

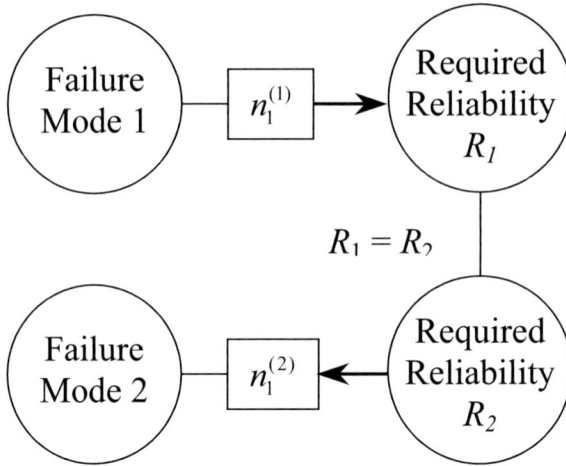

Fig. 3.10 Suggested course of action: equal reliability allocation may justify safety factors associated with different failure modes

Chapter 4

Safety Factors and Reliability: Deterministic Actual Stress & Random Yield Stress

"The strength of a structure cannot be exactly calculated in advance. It is dependent a several quantities, which are accurately known, and therefore more or less uncertain."

A.I. Johnson (1953)

"In deterministic design, the capacity reduction factor and load factor are determined subjectively based on judgement, intuition, and experience; in probabilistic design, they are estimated explicitly project by project considering the specific conditions, giving more control to the design engineers."

A. Haldar and S. Mahadevan (2000)

"Since the late Professor A.M. Freudenthal presented a rational approach to the structural safety…an ever-increasing effort has been directed toward the application of the theory of probability and statistics in structural engineering."

J.T.P. Yao (1985)

"Until all future designs use probabilistic approaches, the structural analysis community must use both deterministic and probabilistic approaches. The "rationale" of these two approaches is quite different, and therefore a temporary process that bridges the two should be implemented."

G. Maymon (2002)

In the previous chapter we studied the case in which the actual stress was treated as a random variable, while the yield stress was considered as a deterministic quantity. In this report we investigate the reverse case, namely, when the actual stress is deterministic, while the yield stress is treated as a random variable. Various probability densities to model the actual behavior of the structural element in question are considered.

4.1 Yields Stress Has an Uniform Probability Density, Actual Stress is Deterministic

Let the yield stress have an uniform probability density

$$f_{\Sigma_y}(\sigma_y) = \begin{cases} \dfrac{1}{\sigma_{y,U} - \sigma_{y,L}} & for \ \ \sigma_{y,L} < \sigma < \sigma_{y,U} \\ 0, \ \ otherwise \end{cases} \tag{4.1}$$

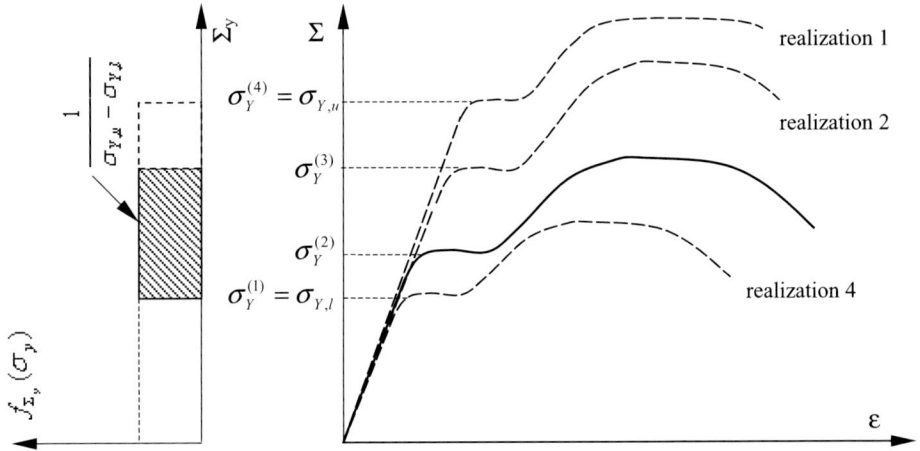

Fig. 4.1 Samples of stress-strain relationships associated with random yield stress

where $\sigma_{y,L}$ is the lower possible level the yield stress may take; $\sigma_{y,U}$ is the upper possible level the yield stress may assume.

The probability distribution function of the yield stress reads

$$F_{\Sigma_y}(\sigma_y) = \begin{cases} 0, \ \ for \ \ \sigma_y < \sigma_{y,L} \\ \dfrac{\sigma_y - \sigma_{y,L}}{\sigma_{y,U} - \sigma_{y,L}} & for \ \ \sigma_{y,L} \le \sigma_y < \sigma_{y,U} \\ 1, \ for \ \ \sigma_{y,U} \le \sigma_y \end{cases} \tag{4.2}$$

The reliability equals

$$R = Prob(\Sigma \le \Sigma_y) = Prob(\sigma \le \Sigma_y)$$
$$= Prob(\Sigma_y \ge \sigma) = 1 - Prob(\Sigma_y \le \sigma) \tag{4.3}$$

Thus, in view Eq. (4.2), we get

$$R = \begin{cases} 1, & for \quad \sigma_y < \sigma_{y,L} \\ 1 - \dfrac{\sigma - \sigma_{y,L}}{\sigma_{y,U} - \sigma_{y,L}} & for \quad \sigma_{y,L} \le \sigma_y < \sigma_{y,U} \\ 0, & for \quad \sigma_{y,U} \le \sigma_y \end{cases} \tag{4.4}$$

Consider the case in which the yield stress belongs to the interval $[\sigma_{y,L}, \sigma_{y,U}]$. In this case from Eq. (4.4) we have for the reliability

$$R = \frac{\sigma_{y,U} - \sigma}{\sigma_{y,U} - \sigma_{y,L}} \tag{4.5}$$

We note that the mean yield stress equals

$$E(\Sigma_y) = \tfrac{1}{2}(\sigma_{y,L} + \sigma_{y,U}) \tag{4.6}$$

whereas the variance of the yield stress reads

$$Var(\Sigma_y) = \tfrac{1}{12}(\sigma_{y,U} - \sigma_{y,L})^2 \tag{4.7}$$

We express upper level of the yield stress $\sigma_{y,U}$ as follows, in terms of the mean yield stress $E(\Sigma_y)$ and variance of the yield stress $Var(\Sigma_y)$ via Eqs. (4.6) and (4.7):

$$\sigma_{y,U} = \frac{2E(\Sigma_y) + \sqrt{12Var(\Sigma_y)}}{2}$$
$$= E(\Sigma_y) + \sqrt{3Var(\Sigma_y)} \tag{4.8}$$

The denominator in Eq. (4.5) is directly expressible by the variance as $2\sqrt{3Var(\Sigma_y)}$ in Eq. (4.7). Thus, the reliability in Eq. (4.5) can be rewritten as

$$R = \frac{E(\Sigma_y) + \sqrt{3Var(\Sigma_y)} - \sigma}{2\sqrt{3Var(\Sigma_y)}} \tag{4.9}$$

We divide both the numerator and denominator by σ and express the ratio

$$\frac{\sqrt{3Var(\Sigma_y)}}{\sigma} = \frac{\sqrt{3Var(\Sigma_y)}}{E(\Sigma_y)} \frac{E(\Sigma_y)}{\sigma}$$
$$= \sqrt{3}v_{\Sigma_y} n_1 \tag{4.10}$$

where

$$v_{\Sigma_y} = \sqrt{Var(\Sigma_y)} / E(\Sigma_y) \tag{4.11}$$

is the coefficient of variation of the yield stress,

$$n_1 = E(\Sigma_y) / \sigma \tag{4.12}$$

is the central safety factor. Reliability reads:

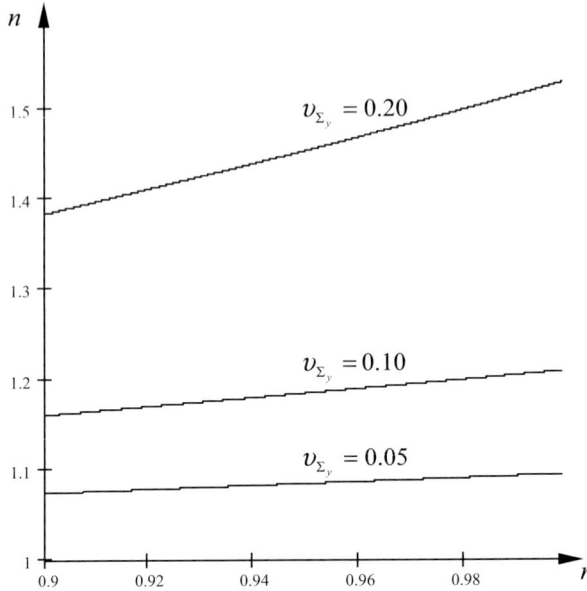

Fig. 4.2 Central safety factor _n_ vs.desired reliability _r_

$$R = \frac{n_1(1+\sqrt{3}v_{\Sigma_y})-1}{2\sqrt{3}v_{\Sigma_y}\,n_1} \qquad (4.13)$$

Eq. (4.13) allows to express the central safety factor as a function of the required reliability _r_:

$$n_1 = \frac{1}{1+\sqrt{3}v_{\Sigma_y}(1-2r)} \qquad (4.14)$$

This equation is remarkable for the required reliability _r_ is *directly* connected with the central *safety factor* n_1. Thus, if the required reliability 0.9 is set, at the coefficient of variation of the yield stress $v_{\Sigma_y} = 0.05$ we get the level of safety factor equal 1.07; for $r = 0.99$ we get $n_1 = 1.09$; reliability level 0.999 corresponds to $n_1 = 1.095$. At the greater coefficient of variation, namely, that comprising 10% we get

$$
\begin{aligned}
n_1 &= 1.16, \quad for \quad r = 0.9 \\
n_1 &= 1.20, \quad for \quad r = 0.99 \\
n_1 &= 1.21, \quad for \quad r = 0.999
\end{aligned}
\qquad (4.15)
$$

When the variability constitutes 20%, we obtain

$$n_1 = 1.38, \quad for \quad r = 0.9$$
$$n_1 = 1.51, \quad for \quad r = 0.99 \tag{4.16}$$
$$n_1 = 1.53, \quad for \quad r = 0.999$$

etc. yielding greater needed safety factors with greater demanded reliability level.

4.2 Yield Stress Has an Exponential Probability Density, Actual Stress Is Deterministic

Consider now the case in which Σ_y is variable with exponential probability density but Σ is deterministic. Hence Σ takes only a single value σ with unity probability.

The probability density of Σ_y reads:

$$f_{\Sigma_y}(\sigma_y) = \begin{cases} 0, & for \quad \sigma_y < 0 \\ a\exp(-a\sigma_y), & for \quad \sigma_y \geq 0 \end{cases} \tag{4.17}$$

Here f_{Σ_y} is the probability density of the yield stress.

The probability distribution function of Σ_y is defined as (Fig. 4.3)

$$F_{\Sigma_y}(\sigma_y) = Prob(\Sigma_y \leq \sigma_y) \tag{4.18}$$

i.e. as a probability that Σ_y will take values that are not in excess of any pre-selected value σ_y.

According to the definition of the probability distribution

$$F_{\Sigma_y}(\sigma_y) = \int_{-\infty}^{\sigma_y} f_{\Sigma_y}(t)dt \tag{4.19}$$

we get

$$F_{\Sigma_y}(\sigma_y) = \begin{cases} 0, & for \quad \sigma_y < 0 \\ 1 - \exp(-a\sigma_y), & for \quad \sigma_y \geq 0 \end{cases} \tag{4.20}$$

The parameter a is the reciprocal of the mathematical expectation

$$E(\Sigma_y) = \frac{1}{a} \tag{4.21}$$

The reliability reads

$$R = Prob(\Sigma \leq \Sigma_y) = Prob(\sigma \leq \Sigma_y)$$
$$= Prob(\Sigma_y \geq \sigma) = 1 - Prob(\Sigma_y \leq \sigma) \tag{4.22}$$

In the right side of the equation (4.22) we recognize that the quantity $Prob(\Sigma_y \leq \sigma)$ coincides with Eq. (4.18) when instead of σ_y in Eq. (4.18) we substitute σ. In other words $Prob(\Sigma_y \leq \sigma)$ equals the probability distribution function of the yield stress evaluated at the level of the actual stress (Fig. 4.4):

$$Prob(\Sigma_y \leq \sigma) = F_{\Sigma_y}(\sigma)$$

(4.23)

Thus, bearing in mind Eq. (4.20) we get

$$Prob(\Sigma_y \leq \sigma) = \begin{cases} 0, & for \quad \sigma \leq 0 \\ 1 - \exp(-a\sigma), & for \quad \sigma \geq 0 \end{cases}$$

(4.24)

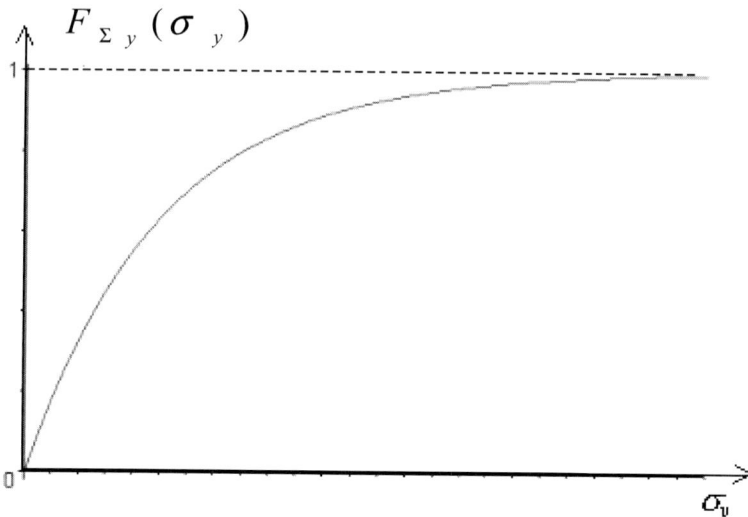

Fig. 4.3 Probability distribution of the yield stress

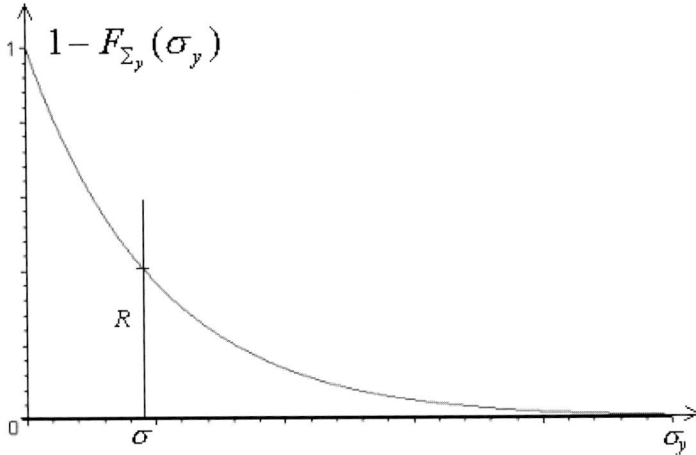

Fig. 4.4 Reliability equals the function $1 - F_{\Sigma_y}(\sigma_y)$ **evaluated at the actual stress**

Hence, the reliability in Eq. (4.22) becomes

$$R = 1 - Prob(\Sigma_y \le \sigma) = 1 - F_{\Sigma_y}(\sigma)$$
$$= 1 - [1 - \exp(-a\sigma)]U(\sigma) \qquad (4.25)$$

where $U(\sigma)$ is the unit step function, *i.e.* $U(\sigma)$ equals unity for positive σ and vanishes otherwise. Taking into account the relationship (4.21) we can rewrite Eq. (4.25) in the following manner

$$R = \left\{ \exp\left[-\frac{\sigma}{E(\Sigma_y)} \right] \right\} U(\sigma) \qquad (4.26)$$

We recognize the argument in Eq. (4.26) to be reciprocal of the safety factor. Three safety factors coincide in this case. Hence we denote them by a single notation s. Thus we get the following relationship:

$$R = \exp(-1/s)U(\sigma) \qquad (4.27)$$

As we see, reliability is intimately connected with the safety factor in the case under consideration. In fact, the safety factor can be expressed directly from Eq. (4.22) for $\sigma \ge 0$:

$$n = -\frac{1}{\ln(R)} = \frac{1}{\ln(\frac{1}{R})} \qquad (4.28)$$

In this particular case if the required reliability equals 0.9 the safety factor $1/ln(0.9)$ is greater than 9!

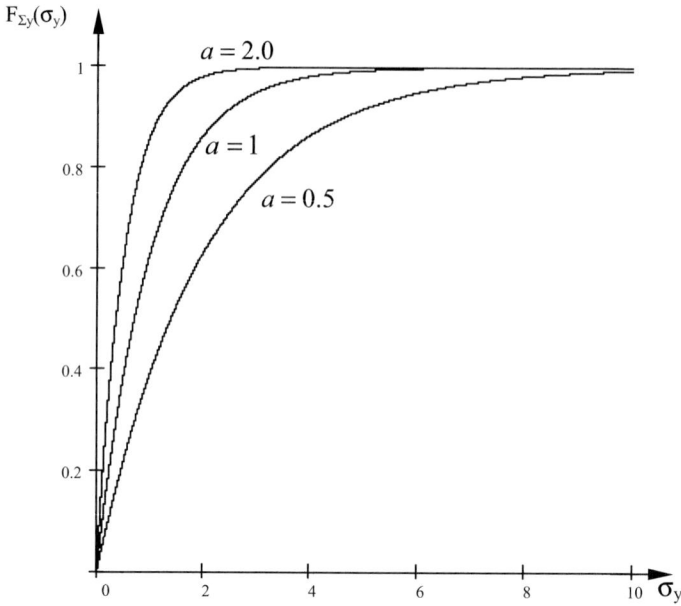

Fig. 4.5 Probability distribution of the yield stress for different values of a

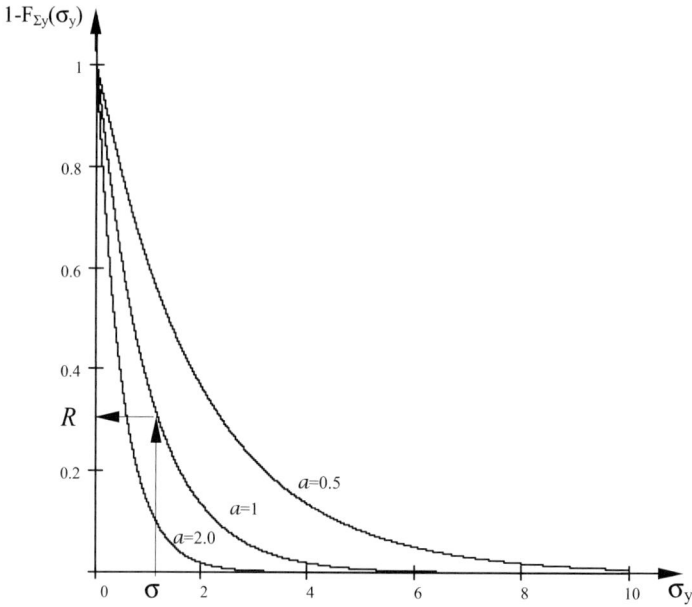

Fig. 4.6 Reliability evaluated at the actual stress level for a=2.0

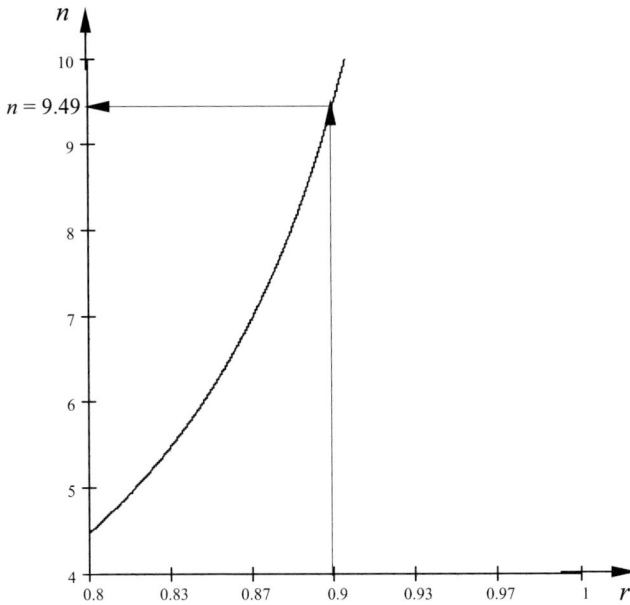

Fig. 4.7 Central safety factor *n* in terms of desired reliability *r*

The results in this case, although may seem to be very surprising, are quite understandable. The variance of the yield stress

$$Var(\Sigma_y) = \frac{1}{a^2} \tag{4.29}$$

The coefficient of variation in this case

$$c.o.v. = \frac{\sqrt{Var(\Sigma_y)}}{E(\Sigma_y)} = \frac{1/a}{1/a} = 1 \tag{4.30}$$

equals unity; *i.e.* there is a large variation around the mean value of the yield stress; hence, large safety factors are needed to achieve the required reliability levels.

4.3   Strength Has a Rayleigh Probability Density, Stress Is Deterministic

The probability density of the strength is given by

$$f_{\Sigma_y}(\sigma_y) = \begin{cases} 0, & for \ \sigma_y < 0 \\ \dfrac{\sigma_y}{b^2} \exp\left(-\dfrac{\sigma_y^2}{2b^2}\right), & for \ \sigma_y \geq 0 \end{cases} \tag{4.31}$$

with parameter b^2. The distribution function is

$$F_{\Sigma_y}(\sigma_y) = \left[1 - \exp\left(-\frac{\sigma_y^2}{2b^2}\right)\right]U(\sigma_y) \tag{4.32}$$

We also note that the mean strength is

$$E(\Sigma_y) = \frac{b\sqrt{\pi}}{\sqrt{2}} \approx 1.25b \tag{4.33}$$

whereas the variance of the strength is

$$Var(\Sigma_y) = \frac{(4-\pi)b^2}{2} \approx 0.43b^2 \tag{4.34}$$

Reliability is given by Eq. (4.32)

$$R = 1 - Prob(\Sigma_y \le \sigma) = 1 - F_{\Sigma_y}(\sigma) \tag{4.35}$$

Bearing in mind Eq. (4.32) we get:

$$R = \left[\exp\left(-\frac{\sigma^2}{2b^2}\right)\right]U(\sigma) \tag{4.36}$$

Now, taking into account Eq. (4.33) we can substitute instead of b,

$$b \approx \frac{E(\Sigma_y)}{1.25} = 0.8E(\Sigma_y) \tag{4.37}$$

to get

$$R = U(\sigma)\exp\left\{-\frac{\sigma^2}{2[0.8E(\Sigma_y)]^2}\right\} \tag{4.38}$$

or, in terms of the central safety factor

$$n = \frac{E(\Sigma_y)}{\sigma} \tag{4.39}$$

we obtain

$$R = U(\sigma)\exp\left\{-\frac{0.78125}{s^2}\right\} \tag{4.40}$$

Safety factor n can be expressed from the reliability

$$n = \sqrt{\frac{0.78125}{\ln(1/R)}} = \frac{0.8839}{\sqrt{\ln(1/R)}} \tag{4.41}$$

Let the reliability be set at $R = 0.99$. Eq. (4.40) yields safety factor 8.82 (see Fig. 4.8). This result is again understandable since the coefficient of variation in this case too is quite large:

$$c.o.v. = \frac{\sqrt{Var(\Sigma_y)}}{E(\Sigma_y)} \approx \frac{\sqrt{0.43b^2}}{1.25b} = 0.52 \tag{4.42}$$

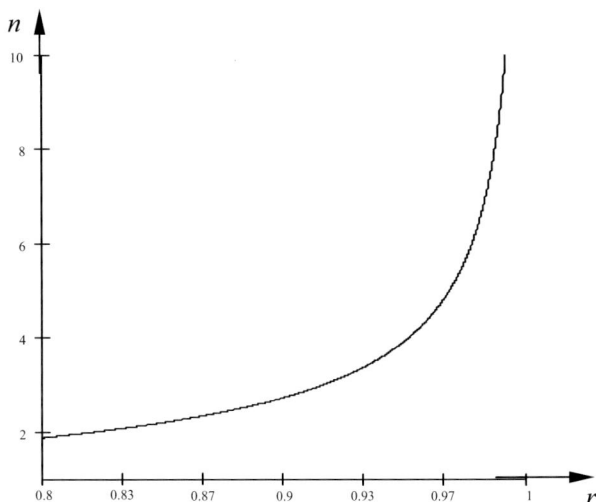

Fig. 4.8 Central safety factor *n*, in terms of desired reliability *r*

4.4 Various Factors of Safety in Buckling

It is best to start with an engineering example, first in the deterministic setting. Consider an element that is simply supported at its ends. It is subjected to the compressive load P at the ends, as well as the concentrate bending moment M. The section modulus (or the moment of resistance as known in European literature) is denoted by W. Material's proportionality limit σ_{pr} as well as the yield stress σ_y are given. We are interested in determining the safety factor in 3 different regimes.

a) Both Concentrated Moment and Axial Force Increase Simultaneously

In this case we have

$$\sigma_{max} = \frac{M_{max}}{W} + \frac{P}{A} \tag{4.43}$$

where M_{max} is the maximum bending moment

$$M_{max} = \frac{M}{\cos\dfrac{kL}{2}} \tag{4.44}$$

where

$$k = \sqrt{\frac{P}{EI}} \tag{4.45}$$

A = cross sectional area. Since the relationship between the stresses and the load P is nonlinear, the safety factor n_{SF} is determined as follows. We multiply the load by n_{SF} so as to achieve a level of stress equal to the yield stress. Thus, the deterministic safety factor is derived from

$$\sigma_y = \frac{n_{SF} M}{W \cos \dfrac{k_y L}{2}} + \frac{n_{SF} P}{A} \tag{4.46}$$

where

$$k_y = \sqrt{\frac{n_{SF} P}{EI}} \tag{4.47}$$

Consider now the probabilistic setting of the problem. Let σ_y be a random variable Σ_y. Then the central safety factor s_1 is determined from the equation, in the manner, analogous to the deterministic setting, except that σ_y is replaced by $E(\Sigma_y)$, and n_{SF} is replaced by s_1:

$$E(\Sigma_y) = \frac{s_1 M}{W \cos \dfrac{k_y L}{2}} + \frac{s_1 P}{A} \tag{4.48}$$

where

$$k_y = \sqrt{\frac{s_1 P}{EI}} \tag{4.49}$$

b) Axial Force Remains Constant, Concentrated Moment Increases

Deterministic safety factor n_{SF} is found from the equation:

$$\sigma_y = \frac{n_{SF} M}{W \cos \dfrac{kL}{2}} + \frac{P}{A} \tag{4.50}$$

whereas the appropriate probabilistic central safety factor is determined from the equation:

$$E(\Sigma_y) = \frac{s_1 M}{W \cos \dfrac{kL}{2}} + \frac{P}{A} \tag{4.51}$$

where

$$k = \sqrt{\frac{P}{EI}} \tag{4.52}$$

c) *Concentrated Moment Remains Constant, Axial Force Increases*

The deterministic safety factor n_{SF} is determined from the equation

$$\sigma_y = \frac{M}{W \cos\dfrac{k_y L}{2}} + \frac{n_{SF} P}{A} \tag{4.53}$$

where

$$k_y = \sqrt{\frac{n_{SF} P}{EI}} \tag{4.54}$$

The probability analog of this equation reads:

$$E(\Sigma_y) = \frac{M}{W \cos\dfrac{k_y L}{2}} + \frac{n_1 P}{A} \tag{4.55}$$

where

$$k_y = \sqrt{\frac{n_1 P}{EI}} \tag{4.56}$$

For example, let $M = 2$ kN·m, $P = 100$ kN. The cross-sectional area is annular with mean diameter $D_m = 10$ cm; thickness $= 0.5$ cm, $L = 3$ m. Then the deterministic safety factors become, in three settings

Case (a): $n_{SF} = 1.85$

Case (b): $n_{SF} = 3.37$

Case (c): $n_{SF} = 2.52$

In probabilistic setting, if $E(\Sigma_y) = 300$ MPa, the same "central" safety factor is obtained. Yet, straightforward application of the definition $E(\Sigma)/\sigma$ would be incorrect, since the load P and the stress σ are not interrelated linearly.

In some cases, a simplified analysis can be performed: Consider the column that is simultaneously subjected to the uniform by distributed load q and axial compressive load P. Then the exact analysis of the differential equation

$$EIw'' + Pw = M_t \tag{4.57}$$

where $w =$ displacement, $M_t =$ bending moment due to the transverse load,

$$M_t = \frac{qL}{2}x - \frac{q}{2}x^2 \tag{4.58}$$

leads to the maximum bending moment

$$M_{max} = \frac{q}{k^2} \cdot \frac{1 - \cos\dfrac{kL}{2}}{\cos\dfrac{kL}{2}} \tag{4.59}$$

where

$$k = \sqrt{P/EI} \qquad (4.60)$$

The approximate relationship, as is well known from the applied theory of elasticity reads

$$M_{max} \approx \frac{qL^2/8}{1 - \dfrac{P}{P_{cr}}} \qquad (4.61)$$

Thus the stresses would read

$$\sigma = \frac{qL^2/8}{W\left(1 - \dfrac{P}{P_{cr}}\right)} + \frac{P}{A} \qquad (4.62)$$

where W is the section modulus. If both q and P increase proportionally, the safety factor is found from equation

$$\sigma_y = \frac{n_{SF}\, qL^2/8}{W\left(1 - \dfrac{n_{SF}P}{P_{cr}}\right)} + \frac{n_{SF}P}{A} \qquad (4.63)$$

The probabilistic "central" safety factor is found from the equation

$$E(\Sigma_y) = \frac{n_1 qL^2/8}{W\left(1 - \dfrac{n_1 P}{P_{cr}}\right)} + \frac{n_1 P}{A} \qquad (4.64)$$

and not what would appear appropriate at the first glance:

$$n_1 = \frac{E(\Sigma_y)}{\dfrac{qL^2/8}{W\left(1 - \dfrac{P}{P_{cr}}\right)} + \dfrac{P}{A}} \qquad (4.65)$$

The correct equation (4.64) leads to a quadratic equation for the central safety factor n_1:

$$n_1^{\,2}\,\frac{P^2 W}{A P_{cr}} - n_1\left(\frac{PE(\Sigma_y)W}{P_{cr}} - \frac{PS}{A}\right) - \frac{qL^2}{8} + E(\Sigma)W = 0 \qquad (4.66)$$

4.5 Yield Stress Has a Weibull Probability Density, Actual Stress Is Deterministic

Consider now the case when the probability distribution function of the yield stress reads

$$F_{\Sigma_y}(\sigma_y) = \exp\left[-\exp\left(\frac{a_{\Sigma_y} - \sigma_y}{b_{\Sigma_y}}\right)\right] \tag{4.67}$$

where a_{Σ_y} and b_{Σ_y} are positive parameters.

Reliability becomes

$$R = Prob(\sigma \le \Sigma_y) = Prob(\Sigma_y \ge \sigma) = 1 - Prob(\Sigma_y \le \sigma)$$
$$= 1 - F_{\Sigma_y}(\sigma) \tag{4.68}$$

thus yielding the reliability as follows:

$$R = 1 - \exp\left[-\exp\left(\frac{a_{\Sigma_y} - \sigma}{b_{\Sigma_y}}\right)\right] \tag{4.69}$$

According to Haldar and Mahadevan (2000) (who do not deal with safety factors in the present context, but discuss the Weibull distribution) the average and variance of Σ_y are directly expressible via a_{Σ_y} and b_{Σ_y}:

$$\frac{1}{b_{\Sigma_y}} = \frac{1}{\sqrt{6}} \frac{\pi}{\sqrt{Var(\Sigma_y)}} \tag{4.70}$$

$$a_{\Sigma_y} = E(\Sigma_y) - 0.5772 b_{\Sigma_y} \tag{4.71}$$

Therefore,

$$E(\Sigma_y) = a_{\Sigma_y} + 0.5772 b_{\Sigma_y} \tag{4.72}$$

$$Var(\Sigma_y) = \frac{\pi^2}{6} b_{\Sigma_y}^2 \tag{4.73}$$

Substitution into Eq. (4.69) yields

$$R = 1 - \exp\left[-\exp\left(\frac{E(\Sigma_y) - 0.5772\frac{\sqrt{6Var(\Sigma_y)}}{\pi} - \sigma}{\sqrt{6Var(\Sigma_y)}/\pi}\right)\right] \tag{4.74}$$

or

$$R = 1 - \exp\left[-\exp\left(\frac{E(\Sigma_y) - 0.45\sqrt{6Var(\Sigma_y)} - \sigma}{0.78\sqrt{6Var(\Sigma_y)}}\right)\right] \tag{4.75}$$

Dividing both the numerator and denominator in Eq. (4.75) by σ, and recalling the definition of central safety factor $n_1 = E(\Sigma_y)/\sigma$ and variability coefficient of the yield stress $v_{\Sigma_y} = \sqrt{Var(\Sigma_y)}/E(\Sigma_y)$ we get

$$R = \exp\left[-\exp\left(\frac{n_1 - 0.45v_{\Sigma_y} - 1}{0.78v_{\Sigma_y} n_1}\right)\right]$$ (4.76)

If the reliability is fixed, one can find the appropriate safety factor:

$$n_1 = \frac{1 + 0.45v_{\Sigma_y}}{1 - 0.78v_{\Sigma_y} \ln(\ln\frac{1}{R})}$$ (4.77)

Since this formula yields safety factors that are less than unity, the use of the Weibull distribution appears to be questionable in the case in question. This leads to an all-important lesson: Direct randomization of the deterministic problem not always may be advisable.

4.6 Yield Stress Has a Fréchet Distribution, and Actual Stress Is Deterministic

The probability distribution function of the yield stress reads in this case

$$F_{\Sigma_y}(\sigma_y) = \exp\left[-\left(\frac{a_{\Sigma_y}}{\sigma_y}\right)^k\right]$$ (4.78)

This expression is directly obtainable from Eq. (3.80) by formal substitution $\Sigma \to \Sigma_y, \sigma \to \sigma_y$. The reliability of the element is given by

$$R = 1 - Prob(\Sigma_y \le \sigma) = 1 - F_{\Sigma_y}(\sigma) = 1 - \exp\left[-\left(\frac{a_{\Sigma_y}}{\sigma}\right)^k\right]$$ (4.79)

The mean of the yield stress is given by

$$E(\Sigma_y) = a_{\Sigma_y} \Gamma\left(1 - \frac{1}{k}\right)$$ (4.80)

where $\Gamma(x)$ is the Gamma function. Bearing in mind Eq. (4.80) the reliability is rewritten

as $$R = 1 - \exp\left[-\left(\frac{E(\Sigma_y)}{\sigma \Gamma\left(1 - \frac{1}{k}\right)}\right)^k\right]$$ (4.81)

or, in view of the definition of the central safety factor n in Eq. (4.12),

$$R = 1 - \exp\left[-\left(\frac{n}{\Gamma\left(1-\frac{1}{k}\right)}\right)^{k}\right] \quad (4.82)$$

The safety factor is directly expressible via the required reliability level $R=r$:

$$n = \Gamma\left(1-\frac{1}{k}\right)\ln\left[\frac{1}{(1-r)}\right]^{\frac{1}{k}} \quad (4.83)$$

Table 4.2 lists values of k, required reliabilities the attendant central safety factor.

Table 4.2. Reliability and Appropriate Safety Factors for Various Values of k

r	k	n	r	n	r	n	r	n
	1.5	4.671		5.567		7.415		9.717
	2.0	2.690		3.068		3.804		4.658
0.9	3.0	1.788	0.95	1.952	0.99	2.253	0.999	2.579
	4.0	1.510		1.612		1.795		1.987
	5.0	1.376		1.450		1.580		1.714

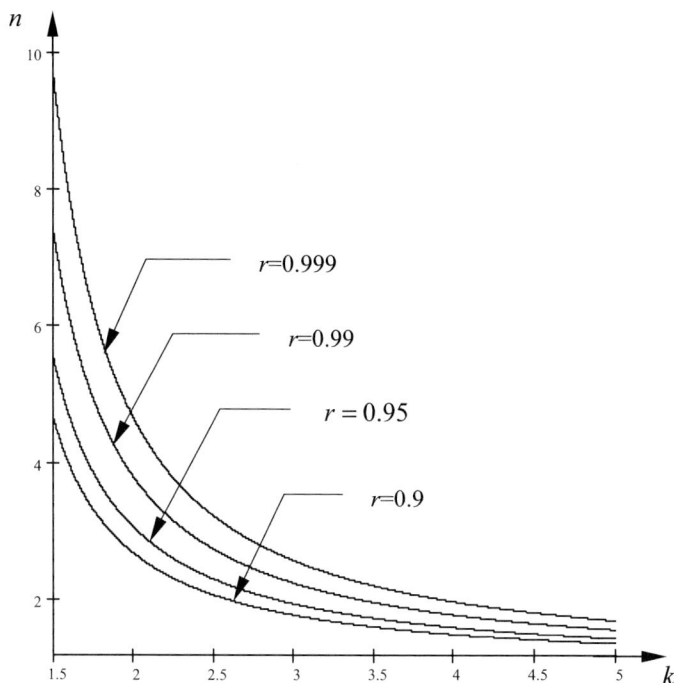

Fig. 4.9 Central safety factor n vs. coefficient k, for various required reliabilities

4.7   Yield Stress has a Three Parameter Weibull Distribution, and Actual Stress Is Deterministic

In this case the probability distribution function reads

$$F_{\Sigma_y}(\sigma_y) = 1 - \exp\left[-\left(\frac{\sigma - \sigma_0}{a_{\Sigma_y} - \sigma_0}\right)^k\right] \tag{4.84}$$

where σ_0 is the minimum possible value of $\Sigma_y \geq 0$; a_{Σ_y} is the characteristic parameter ($a_{\Sigma_y} \geq \sigma_0$), and k is the positive shape parameter. The reliability R equals,

$$R = 1 - Prob(\Sigma_y \leq \sigma) = 1 - F_{\Sigma_y}(\sigma) = 1 - \left[1 - \exp\left[-\left(\frac{\sigma - \sigma_0}{a_{\Sigma_y} - \sigma_0}\right)^k\right]\right] =$$

$$= \exp\left[-\left(\frac{\sigma - \sigma_0}{a_{\Sigma_y} - \sigma_0}\right)^k\right] \tag{4.85}$$

The mean value $E(\Sigma_y)$ of the yield stress reads

$$E(\Sigma_y) = \sigma_0 + (a_{\Sigma_y} - \sigma_0)\Gamma\left(1 + \frac{1}{k}\right) \tag{4.86}$$

The reliability in terms of the mean yield stress becomes,

$$R = \exp\left\{-\left[\frac{(\sigma - \sigma_0)\Gamma\left(1 + \frac{1}{k}\right)}{E(\Sigma_y) - \sigma_0}\right]^k\right\} \tag{4.87}$$

Thus the central safety factor $n = E(\Sigma_Y)/\sigma$ in terms of required reliability r, is given by the following formula

$$n = \delta + \frac{(1-\delta)\Gamma\left(1 + \frac{1}{k}\right)}{\left[\ln\left(\frac{1}{r}\right)\right]^{\frac{1}{k}}} \tag{4.88}$$

where $\delta = \sigma_0/\sigma$, and $0 < \delta < 1$. Table 4.3 lists values of central safety factor in terms of shape parameter k, and required reliabilities r, while δ is fixed at 0.9. The greater the dispersion between σ_0 and σ ($\delta \rightarrow 0$), the greater the safety factor is required.

Table 4.2: Reliability and Corresponding Safety Factors for Various Values of k

r	k	n	r	n	r	n	r	n
	1.5	1.305		1.554		2.838		9.914
	2.0	1.173		1.291		1.784		3.702
0.9	3.0	1.089	0.95	1.140	0.99	1.314	0.999	1.793
	4.0	1.059		1.090		1.186		1.410
	5.0	1.044		1.066		1.130		1.265

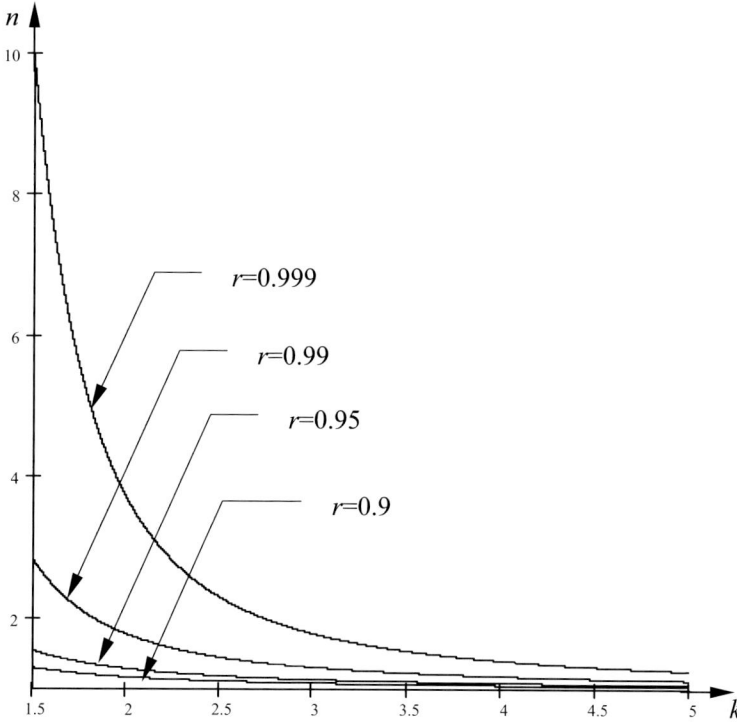

Fig. 4.10 Central safety factor vs. coefficient k for various required reliabilities

4.8   Yield Stress Has a Two Parameter Weibull Distribution, and Actual Stress is Deterministic

The probability distribution function Σ_y reads

$$F_{\Sigma_y}(\sigma_y) = 1 - \exp\left[-\left(\frac{\sigma_y}{a_{\Sigma_y}}\right)^k\right], \quad \sigma_0 \geq 0 \tag{4.89}$$

The reliability R becomes,

$$R = 1 - Prob(\Sigma_y \leq \sigma) = 1 - F_{\Sigma_y}(\sigma) = 1 - \left[1 - \exp\left[-\left(\frac{\sigma}{a_{\Sigma_y}}\right)^k\right]\right] =$$

$$= \exp\left[-\left(\frac{\sigma}{a_{\Sigma_y}}\right)^k\right], \tag{4.90}$$

while the mean value $E(\Sigma_y)$ of the yield stress equals

$$E(\Sigma_y) = a_{\Sigma_y}\,\Gamma\left(1 + \frac{1}{k}\right) \tag{4.91}$$

In view of Eq. (4.91), Eq. (4.90) can be rewritten as follows

$$R = \exp\left\{-\left[\frac{\sigma\,\Gamma\left(1 + \dfrac{1}{k}\right)}{E(\Sigma_y)}\right]^k\right\} \tag{4.92}$$

Bearing in mind Eq. (4.12) we put Eq. (4.87) as follows:

$$R = \exp\left\{-\left[\frac{\Gamma\left(1 + \dfrac{1}{k}\right)}{n}\right]^k\right\} \tag{4.93}$$

The central safety factor n that corresponds to the required reliability r is obtained as follows:

$$n = \frac{\Gamma\left(1 + \dfrac{1}{k}\right)}{\left[\ln\left(\dfrac{1}{r}\right)\right]^{\frac{1}{k}}} \tag{4.94}$$

Table 4.4 lists values of central safety factor n in terms of k, and required reliabilities r.

Table 4.4: Reliability and Central Safety Factors for various Values of k

r	k	n	r	n	r	n	r	n
0.9	2.0	2.73	0.95	3.91	0.99	8.84	0.999	28.02
	3.0	1.89		2.40		4.13		8.93
	4.0	1.59		1.90		2.86		5.10
	5.0	1.44		1.66		2.30		3.66

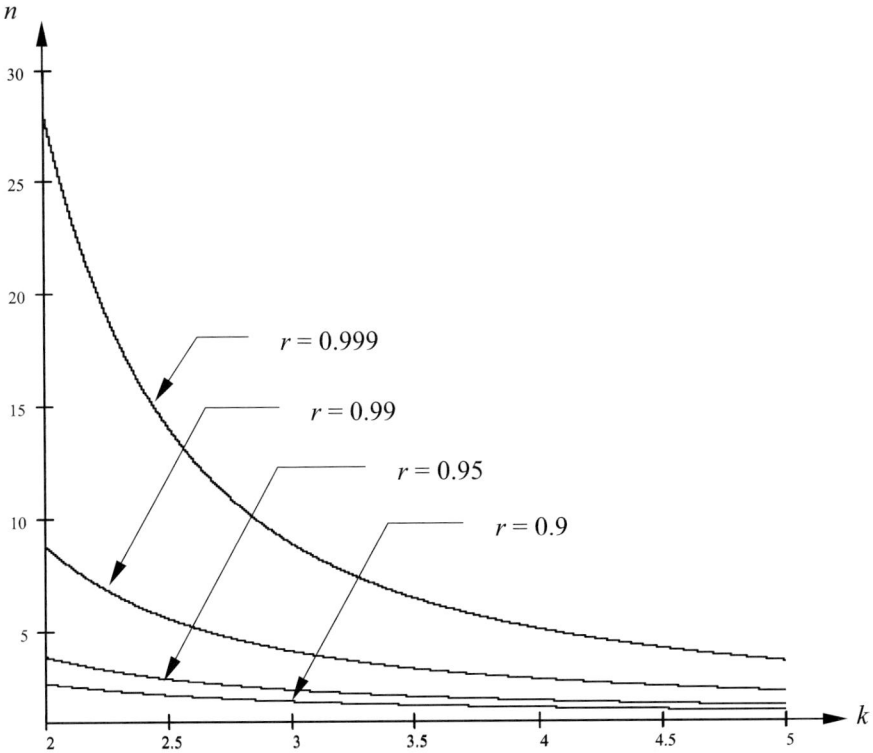

Fig. 4.11 Central safety factor n vs. coefficient k, for various required reliabilities

4.9 Concluding Comments on Proper Distribution Functions

In this chapter we dealt with the case that is reverse to that discussed in Chapter 3. Bolotin (1968) notes about the probability distribution function appropriate to describe the random variable involved:

"For calculations usually the most comfortable is the *normal distribution*; however, its use for describing the strength characteristics is not correct due to the fact that the distribution takes place also on the negative semi-axis. It appears more justified to use *log-normal distribution, Rayleigh distribution, Weibull distribution.*"

As Mischke (1970) notes, "there is disenchantment with the term factor of safety." In order to replace it with reliability calculations, one needs the information on the distribution functions of the strength. On this subject, Shigley (1970) writes: "One of the unfortunate facts of life is that there are almost no publications of data on the distribution of stress and strength."

Whereas our interest lies in the case where an element is subjected to a random (or deterministic) stress, and is made of the material whose yield stress is deterministic (or random), the results that will be reported, are applicable to numerous other situations. First of all, the material may not possess the yield stress. Then the ultimate stress, or other dangerous level of stress must replace the yield stress concept. In their review article, Bhattacharyya and Johnson (1981) list other pertinent cases:

> "Example 1. [Rocket engines] Let X represent maximum chamber pressure generated by idnition of a solid propellant, and Y be the strength of the rocket chamber. Then $R=Prob\ (Y>X)$ is the probability of a successful firing of the engine.
>
> Example 2. [Comparing two treatments] A standard design for the comparison of two drugs is to assign drug A to one group of subjects and drug B to another group. Denote by X and Y the remission times with drug A and B, respectively. Inferences about $R=Pr(Y>X)$ based on the remission times $X_1, X_2, .., Y_m$, are of primary interest to the experimenter. Although the name 'stress-strength' is not appropriate in the present context, our target of inference is the parameter R which has the same structure as in Example 1."

Likewise, O'Connor (1985) notes:

> "Load and strength are considered in a widest sense. 'Load' might refer to a mechanical stress, a voltage or internally generated stresses such as temperature. 'Strength' might refer to any resisting property, such as hardness, strength, melting point or adhesion. Examples are:
> 1.   A bearing fails when the internally generated loads (due to perhaps to roughness, loss of lubricity, etc.) exceed the load strength causing fracture, overheating or seizure.
> 2.   A transistor gate in an integrated circuit fails when the voltage applied causes a local current density, and hence temperature rise, above the melting point of the conductor or semiconductor material.
> 3.   A hydraulic valve fails when the seal cannot withstand the applied pressure without leaking excessively.
> 4.   A shaft fractures when torque exceeds strength."

Chapter 5

Safety Factor and Reliability:
Both Actual Stress and Yield Stress Are Random

"In the general case, when both the loads as well as the characteristic strength are random, allocation of normative safety factor is quite a difficult problem, demanding the initial resolution of the reliability problem."

V.V. Bolotin (1968)

"Many believe that use of a safety factor smaller than some preconceived magnitude (e.g. 2.5) will result in the failure. Actually, with much high safety factors, the failure probability may vary from a satisfactory level to an intolerable high. A safety factor of one implies, to many that failure will occur 100% of the time, because there is no safety margin. Actually, if strength and stress are normally distributed, failure will occur 50% of the time. It is well known that distributions exist in both available capacity and demand requirements. It is these distributions (as defined by mean values, standard deviations and other parameters) with which the designer should be concerned. The safety factor concepts overlook the factor of variability, which may give different probabilities of failure for any one given safety factor."

C.O. Smith (1980)

"Generally, failures can be viewed as overstress."

M. Moderres (1993)

In this chapter both stress and strength are treated as random variables. Thus, this is the most general case within the context of random variables theory. The connection between the reliability and safety factors will be demonstrated for various possible distributions.

5.1 Introductory Comments

In this part we consider most realistic case when both the yield stress Σ_y and the actual stress Σ are represented as random variables. The reliability reads:

$$R = Prob(\Sigma \leq \Sigma_y) \tag{5.1}$$

We denote by $f_{\Sigma\Sigma_y}(\sigma, \sigma_y)$ a joint probability density function of Σ and Σ_y. Then Eq. (5.1) becomes

$$R = \iint_{\Sigma \leq \Sigma_y} f_{\Sigma\Sigma_y}(\sigma, \sigma_y) d\sigma d\sigma_y = \int_0^\infty \left[\int_\sigma^\infty f_{\Sigma\Sigma_y}(\sigma, \sigma_y) d\sigma_y \right] d\sigma \tag{5.2}$$

or, alternatively,

$$R = \iint_{\Sigma \le \Sigma_y} f_{\Sigma\Sigma_y}(\sigma,\sigma_y)d\sigma d\sigma_y = \int_0^\infty \left[\int_0^{\sigma_y} f_{\Sigma\Sigma_y}(\sigma,\sigma_y)d\sigma \right] d\sigma_y \qquad (5.3)$$

where the integration domain extends over the region in which Σ and Σ_y to be independent random variables. We find two formulas, stemming from Eqs (5.2) and (5.3), respectively:

$$R = \int_0^\infty [1 - F_{\Sigma_y}(\sigma)] f_\Sigma(\sigma) d\sigma \qquad (5.4)$$

$$R = \int_0^\infty F_\Sigma(\sigma_y) f_{\Sigma_y}(\sigma_y) d\sigma_y \qquad (5.5)$$

We will use either of Eqs. (5.4) or (5.5) based on convenience of computation.

5.2   Both Actual Stress and Yield Stress Have Normal Probability Density

Let

$$f_\Sigma(\sigma) = \frac{1}{b_\Sigma \sqrt{2\pi}} \exp\left[-\frac{1}{2}\left(\frac{\sigma - E(\Sigma)}{b_\Sigma} \right)^2 \right] \qquad (5.6)$$

$$f_{\Sigma_y}(\sigma_y) = \frac{1}{b_{\Sigma_y} \sqrt{2\pi}} \exp\left[-\frac{1}{2}\left(\frac{\sigma_y - E(\Sigma_y)}{b_{\Sigma_y}} \right)^2 \right] \qquad (5.7)$$

where

$E(\Sigma)$ = mean value of the actual stress

$b_\Sigma = \sqrt{Var(\Sigma)}$ = standard deviation of the actual stress

$E(\Sigma_y)$ = mean value of the yield stress

$b_{\Sigma_y} = \sqrt{Var(\Sigma_y)}$ = standard deviation of the yield stress

We introduce a new random variable

$$M = \Sigma_y - \Sigma \qquad (5.8)$$

which is called the safety margin. Since Eq. (5.8) expresses linearly Σ and Σ_y, the safety margin, as a linear function of the normal variables, is also a normal variable with the mean value

$$E(M) = E(\Sigma_y) - E(\Sigma) \qquad (5.9)$$

and standard deviation b_M found as follows

$$b_M = \sqrt{b_\Sigma^2 + b_{\Sigma_y}^2} \qquad (5.10)$$

The reliability is then

$$R = Prob(\Sigma \leq \Sigma_y) = Prob(M \geq 0)$$

$$= \int_0^\infty \frac{1}{b_M \sqrt{2\pi}} exp\left[-\left(\frac{t - E(M)}{b_M}\right)^2\right] dt \qquad (5.11)$$

To perform an integration in Eq. (5.11), we introduce new variable

$$z = \frac{t - E(M)}{b_M} \qquad (5.12)$$

Hence

$$dt = b_M dz \qquad (5.13)$$

Also, when $t = 0$, the lower limit of z equals

$$z = \frac{0 - E(M)}{b_M} = -\frac{E(\Sigma_y) - E(\Sigma)}{\sqrt{b_\Sigma^2 + b_{\Sigma_y}^2}} \qquad (5.14)$$

Hence,

$$R = \frac{1}{\sqrt{2\pi}} \int_{-\frac{E(\Sigma_y) - E(\Sigma)}{\sqrt{b_\Sigma^2 + b_{\Sigma_y}^2}}}^{\infty} exp(-z^2/2) dz \qquad (5.15)$$

Thus, reliability in Eq. (5.11) becomes

$$R = 1 - \frac{1}{\sqrt{2\pi}} \int_{-\infty}^{\frac{E(\Sigma_y) - E(\Sigma)}{\sqrt{b_\Sigma^2 + b_{\Sigma_y}^2}}} exp(-z^2/2) dz$$

$$= 1 - \Phi\left(-\frac{E(\Sigma_y) - E(\Sigma)}{\sqrt{b_\Sigma^2 + b_{\Sigma_y}^2}}\right) \qquad (5.16)$$

This formula can be rewritten in several alternative ways. First of all we note that

$$\frac{E(\Sigma_y) - E(\Sigma)}{\sqrt{b_\Sigma^2 + b_{\Sigma_y}^2}} = \frac{1}{v_M} \qquad (5.17)$$

where v_M is the coefficient of variation of the safety margin. Thus

$$R = 1 - \Phi\left(-\frac{1}{v_M}\right) \qquad (5.18)$$

We also introduce coefficients of variation of the actual stress and the yield stress, respectively,

$$v_{\Sigma} = \frac{b_{\Sigma}}{E(\Sigma)} \tag{5.19}$$

$$v_{\Sigma_y} = \frac{b_{\Sigma_y}}{E(\Sigma_y)} \tag{5.20}$$

Then,

$$\sqrt{b_{\Sigma}^2 + b_{\Sigma_y}^2} = E(\Sigma)\sqrt{\frac{b_{\Sigma}^2}{E^2(\Sigma)} + \frac{b_{\Sigma_y}^2}{E^2(\Sigma)}} = E(\Sigma)\sqrt{v_{\Sigma}^2 + \frac{b_{\Sigma_y}^2}{E^2(\Sigma_y)}\frac{E^2(\Sigma_y)}{E^2(\Sigma)}}$$

$$= E(\Sigma)\sqrt{v_{\Sigma}^2 + v_{\Sigma_y}^2 n_1^2} \tag{5.21}$$

Hence, the reliability in Eq. (5.16) becomes

$$R = 1 - \Phi\left(\frac{1 - n_1}{\sqrt{v_{\Sigma}^2 + v_{\Sigma_y}^2 n_1^2}}\right) \tag{5.22}$$

As is seen the reliability R, the central safety factor n_1 and the variabilities v_{Σ}^2 and $v_{\Sigma_y}^2$ are directly interrelated. Eq. (5.22) can be rewritten in a different form to answer the following question: What is the required safety factor n_1 for any specified reliability r? For the specified required reliability r we first find the argument in Eq. (5.22) $(1 - n_1)/\sqrt{v_{\Sigma}^2 + v_{\Sigma_y}^2 n_1^2} = \alpha_r$, from the standard normal tables. This yields an equation

$$n_1^2\left(1 - \alpha_r^2 v_{\Sigma_y}^2\right) - 2n_1 + 1 - \alpha_r^2 v_{\Sigma}^2 = 0 \tag{5.23}$$

yielding

$$n_1 = \left[1 \pm \sqrt{1 - (1 - \alpha_r^2 v_{\Sigma_y}^2)(1 - \alpha_r^2 v_{\Sigma}^2)}\right]/(1 - \alpha_r^2 v_{\Sigma_y}^2) \tag{5.24}$$

Consider an example due to Rao (1992). Let $E(\Sigma_y) = 15{,}000 \, \text{lb/in}^2$, $\sigma_{\Sigma_y} = 3{,}000 \, \text{lb/in}^2$, $E(\Sigma) = 10{,}000 \, \text{lb/in}^2$, $\sigma_{\Sigma} = 1{,}000 \, \text{lb/in}^2$. The central factor of safety equals $n_1 = E(\Sigma_y)/E(\Sigma) = 15{,}000/10{,}000 = 1.5$. The coefficients of variation equal $v_{\Sigma} = 1{,}000/10{,}000 = 0.1$; $v_{\Sigma_y} = 3{,}000/15{,}000 = 0.2$. The reliability is given by Eq. (5.22)

$$R = 1 - \Phi\left[-(1.5 - 1)/\sqrt{0.1^2 + 0.4^4 \times 1.5}\right] = 0.94305 \tag{5.25}$$

Formula (5.22) was apparently first derived by Freudenthal (1947). In the same year, it was reported also by Rzhanitsyn (1947). As the footnote (p. 14), Rzhanitsyn notes: "In the Journal *Proceedings* No. 8, 1945 Freudenthal published a method of determining the safety factor, that is externally similar [to that of Rzhanitsyn], but it is not completely correct", relating safety factor with reliability. Present author was unable to locate the *Proceedings* No. 8, 1945, and the above-mentioned paper by Freudenthal. This list of publications by A. M. Freudenthal that is reproduced in the 1976 issue of the journal

Engineering Fracture Mechanics, does not list the above paper. The earliest paper by Freudenthal (1938) on the probabilistic mechanics whose title page is reproduced on p. 15 of this book does not relate directly safety factor with reliability. Freudenthal's earliest paper that accomplishes this task is dated 1947. Therefore, Rzhanitsyn's (1947) statement on the "not completely correct" formula by Freudenthal cannot be validated. Morever, one can state that the above formula was not derived in the book by Streletskii (1947) or earlier authors (although this writer did not yet succeed to obtain the copy of Streletskii's book). However, Streletskii's book *Selected Works* (1975) reproduces excerpts from his (1947) monograph. There direct connection between reliability and safety factor is not found. Indeed, Rzhanitsyn (1978, p. 5) mentions that "N.S. Streletskii did not succeed to obtain mathematically correct solutions of the connection between the safety factor and [probabilistic] distribution curves of loads and strengths. Nevertheless, his works played an important role in posing the problem of reliability of civil structures, as well as on propagation of statistical methods of analysis." Based on the above evidence, we conclude, that Freudenthal (1947) pioneered this formula.

5.3 Actual Stress Has an Exponential Density, Yield Stress Has a Normal Probability Density

For the titled case the probability densities of Σ and Σ_y read, respectively,

$$f_\Sigma(\sigma) = a\exp(-a\sigma), \quad for \quad \sigma \geq 0$$

$$f_{\Sigma_y}(\sigma_y) = \frac{1}{b_{\Sigma_y}\sqrt{2\pi}}\exp\left[-\frac{1}{2}\left(\frac{\sigma_y - E(\sigma_y)}{b_{\Sigma_y}}\right)^2\right], \quad for \quad -\infty \leq \sigma_y \leq \infty \tag{5.26}$$

We note that

$$E(\Sigma) = \frac{1}{a}, \quad Var(\Sigma) = \frac{1}{a^2} \tag{5.27}$$

$$E(\Sigma_y) = b_{\Sigma_y}, \quad Var(\Sigma_y) = b_{\Sigma_y}^{\,2}$$

We evaluate reliability function as follows

$$R = \int_0^\infty f_{\Sigma_y}(\sigma_y)\left[\int_0^{\sigma_y} f_\Sigma(\sigma)d\sigma\right]d\sigma_y \tag{5.28}$$

Now, the inner integral equals:

$$\int_0^{\sigma_y} f_\Sigma(\sigma)d\sigma = \int_0^{\sigma_y} a\exp(-a\sigma)d\sigma$$

$$= 1 - \exp(-a\sigma_y) \tag{5.29}$$

This results in the following evaluation:

$$R = \int_0^\infty \frac{1}{b_{\Sigma_y}\sqrt{2\pi}} \exp\left[-\frac{1}{2}\left(\frac{\sigma_y - E(\Sigma_y)}{b_{\Sigma_y}}\right)^2\right](1 - e^{-a\sigma_y})d\sigma_y$$

$$= \frac{1}{b_{\Sigma_y}\sqrt{2\pi}} \int_0^\infty \exp\left[-\frac{1}{2}\left(\frac{\sigma_y - E(\Sigma_y)}{b_{\Sigma_y}}\right)^2\right]d\sigma_y \qquad (5.30)$$

$$- \frac{1}{b_{\Sigma_y}\sqrt{2\pi}} \int_0^\infty \exp\left[-\frac{1}{2}\left(\frac{\sigma_y - E(\Sigma_y)}{b_{\Sigma_y}}\right)^2\right]e^{-a\sigma_y}d\sigma_y$$

$$= 1 - \Phi\left(-\frac{E(\Sigma_y)}{b_{\Sigma_y}}\right) - \frac{1}{b_{\Sigma_y}\sqrt{2\pi}} \int_0^\infty \exp\left\{-\frac{1}{2b_{\Sigma_y}^2}\left[\sigma_y - E(\Sigma_y) + ab_{\Sigma_y}^2 + 2E(\Sigma_y)b_{\Sigma_y}^2 - a^2b_{\Sigma_y}^4\right]\right\}d\sigma_y$$

We introduce the following variables

$$t = \frac{\sigma_y - E(\Sigma_y) + ab_{\Sigma_y}^2}{b_{\Sigma_y}} \qquad (5.31)$$

$$b_{\Sigma_y} dt = d\sigma_y$$

The expression for the reliability reads

$$R = 1 - \Phi\left[-\frac{E(\Sigma_y)}{b_{\Sigma_y}}\right] - \frac{1}{\sqrt{2\pi}} \int_{\frac{E(\Sigma_y)-ab_{\Sigma_y}^2}{b_{\Sigma_y}}}^\infty I(t)dt \qquad (5.32)$$

$$I(t) = \exp\left(-\frac{t^2}{2}\right)\exp\left[-\frac{1}{2}\left(2E(\Sigma_y)a - a^2b_{\Sigma_y}^2\right)\right]$$

leading to the final expression

$$R = 1 - \Phi\left[-\frac{E(\Sigma_y)}{b_{\Sigma_y}}\right] - \exp\left[-\frac{1}{2}(2E(\Sigma_y)a - a^2b_{\Sigma_y}^2)\right]\left[1 - \Phi\left(-\frac{E(\Sigma_y) - ab_{\Sigma_y}^2}{b_{\Sigma_y}}\right)\right] \qquad (5.33)$$

This expression is rewritten as follows with notation:

$$v_{\Sigma_y} = \frac{b_{\Sigma_y}}{E(\Sigma_y)}$$

$$\exp\left[-\frac{1}{2}\left(2E(\Sigma_y)a - a^2 b_{\Sigma_y}{}^2\right)\right] = \exp\left[-\frac{1}{2}\left(\frac{E(\Sigma_y)}{E(\Sigma)} - \frac{Var(\Sigma_y)}{E^2(\Sigma)}\right)\right]$$

$$= \exp\left[-\frac{1}{2}\left(n_1 - \frac{Var(\Sigma_y)}{E^2(\Sigma_y)}\frac{E^2(\Sigma_y)}{E^2(\Sigma)}\right)\right] \tag{5.34}$$

$$= \exp\left[-\frac{1}{2}\left(n_1 - v_{\Sigma_y}{}^2 n_1{}^2\right)\right]$$

Also, the expression in Eq. (5.32) becomes

$$1 - \Phi\left[-\frac{E(\Sigma_y) - ab_{\Sigma_y}{}^2}{b_{\Sigma_y}}\right] = 1 - \Phi\left[-\frac{E(\Sigma_y)}{\sqrt{Var(\Sigma_y)}} - \frac{\sqrt{Var(\Sigma_y)}}{E(\Sigma)}\right]$$

$$= 1 - \Phi\left(-\frac{1}{v_{\Sigma_y}} - \frac{\sqrt{Var(\Sigma_y)}}{E(\Sigma_y)}\frac{E(\Sigma_y)}{E(\Sigma)}\right) \tag{5.35}$$

$$= 1 - \Phi\left(-\frac{1}{v_{\Sigma_y}} - v_{\Sigma_y} n_1\right)$$

Thus, all parameters in Eq. (5.32) are expressed in terms of the coefficients of variation and the central safety factor n_1

$$R = 1 - \Phi\left(\frac{1}{v_{\Sigma_y}}\right) - \exp\left[-\frac{1}{2}\left(n_1 - v_{\Sigma_y}{}^2 n_1{}^2\right)\right]\left[1 - \Phi\left(-\frac{1}{v_{\Sigma_y}} - v_{\Sigma_y} n_1\right)\right] \tag{5.36}$$

Consider an example. Let the yield stress has a normal probability density with mean yield stress equals $E(\Sigma_y) = 100$ MPa. The variance equals $Var(\Sigma_y) = 100(MPa)^2$, or standard deviation equals $b_{\Sigma_y} = 10$ MPa, leading to the coefficient of variation to be

$$v_{\Sigma_y} = \frac{b_{\Sigma_y}}{E(\Sigma_y)} = \frac{10}{100} = 0.1 \tag{5.37}$$

The central safety factor n_1 is set at 2, i.e.

$$n_1 = \frac{E(\Sigma_y)}{E(\Sigma)} = 2 \tag{5.38}$$

leading to the value of the mean stress to be equal

$$E(\Sigma) = \frac{1}{2}E(\Sigma_y) = \frac{1}{2} \times 100 = 50 \text{ MPa} \tag{5.39}$$

i.e.

$$a = 1/E(\Sigma) = 1/50 \text{ (MPa)}^{-1} \tag{5.40}$$

Calculations in accordance to the formula (5.36) yield the reliability $R = 0.86194$.

5.4 Actual Stress Has a Normal Probability Density, Strength Has an Exponential Probability Density

In this case we get

$$R = \int_{-\infty}^{\infty} f_\Sigma(\sigma) \left[\int_{\sigma}^{\infty} f_{\Sigma_y}(\sigma_y) d\sigma_y \right] d\sigma \tag{5.41}$$

In new circumstances the probability densities read:

$$f_\Sigma(\sigma) = \frac{1}{\sqrt{2\pi Var(\Sigma)}} \exp\left[-\frac{1}{2}\left(\frac{\sigma - E(\Sigma)}{\sqrt{Var(\Sigma)}} \right)^2 \right]$$

$$f_{\Sigma_y}(\sigma_y) = \frac{1}{E(\Sigma_y)} \exp\left[-\frac{\sigma_y}{E(\Sigma_y)} \right], \quad \sigma_y > 0 \tag{5.42}$$

The reliability becomes:

$$R = \int_{-\infty}^{\infty} \frac{1}{\sqrt{2\pi Var(\Sigma)}} \exp\left[-\left(\frac{\sigma - E(\Sigma)}{2Var(\Sigma)} \right)^2 \right] \exp\left(-\frac{\sigma}{E(\Sigma_y)} \right) d\sigma \tag{5.43}$$

The find expression is as follows:

$$R = \Phi\left(-\frac{E(\Sigma)}{\sqrt{Var(\Sigma)}} \right) + \exp\left[-\frac{1}{2}\left(\frac{2E(\Sigma)}{E(\Sigma_y)} - \frac{Var(\Sigma)}{E^2(\Sigma_y)} \right) \right] \left[1 - \Phi\left(-\frac{E(\Sigma) - Var(\Sigma)/E^2(\Sigma_y)}{\sqrt{Var(\Sigma)}} \right) \right] \tag{5.44}$$

In terms of coefficients of variation and the central safety factor this expression becomes:

$$R = \Phi\left(-\frac{1}{v_\Sigma} \right) + \exp\left[-\frac{1}{2}\left(\frac{2}{n_1} - \frac{v_\Sigma^2}{n_1^2} \right) \right] \left[1 - \Phi\left(-\frac{1}{v_\Sigma} - \frac{v_\Sigma}{n_1^2} \right) \right] \tag{5.45}$$

5.5 Both Actual Stress and Yield Stress Have Log-Normal Probability Densities

Let the actual stress have the following density

$$f_\Sigma(\sigma) = \frac{1}{\sigma b_\Sigma \sqrt{2\pi}} \exp\left[-\frac{(\ln\sigma - a_\Sigma)^2}{2b_\Sigma^2} \right], \quad \sigma > 0 \tag{5.46}$$

where parameters a_Σ and b_Σ are related to the mean value $E(\Sigma)$ as follows:

$$E(\Sigma) = \exp(a_\Sigma + \tfrac{1}{2}b_\Sigma^2) \tag{5.47}$$

The variance of the stress equals

$$Var(\Sigma) = \exp(2a_\Sigma + b_\Sigma^2)[\exp(b_\Sigma^2) - 1] \tag{5.48}$$

The probability density of the yield stress reads:

$$f_{\Sigma_y}(\sigma_y) = \frac{1}{\sigma_y b_{\Sigma_y} \sqrt{2\pi}} \exp\left[-\frac{(\ln \sigma_y - a_{\Sigma_y})^2}{2b_{\Sigma_y}^{\,2}}\right], \quad \sigma_y > 0 \qquad (5.49)$$

The parameters a_{Σ_y} and b_{Σ_y} are related in the following way with the mean yield stress:

$$E(\Sigma_y) = \exp(a_{\Sigma_y} + \tfrac{1}{2} b_{\Sigma_y}^{\,2}) \qquad (5.50)$$

Variance $Var(\Sigma_y)$ is expressed as

$$Var(\Sigma_y) = \exp(2a_{\Sigma_y} + b_{\Sigma_y}^{\,2})[\exp(b_{\Sigma_y}^{\,2}) - 1] \qquad (5.51)$$

The reliability reads

$$R = Prob(\Sigma \le \Sigma_y) \qquad (5.52)$$

Yet, it is easier to express reliability as follows:

$$R = Prob\left(\frac{\Sigma}{\Sigma_y} \le 1\right) \qquad (5.53)$$

Introducing a new variable Z,

$$Z = \frac{\Sigma}{\Sigma_y} \qquad (5.54)$$

we get

$$R = Prob(Z \le 1) \qquad (5.55)$$

which can be rewritten as follows:

$$R = Prob(\ln Z \le 0) \qquad (5.56)$$

We note that

$$\ln Z = \ln \Sigma - \ln \Sigma_y \qquad (5.57)$$

But $\ln \Sigma$ has a normal probability density with

$$E(\ln \Sigma) = a_\Sigma \qquad (5.58)$$

$$Var(\ln \Sigma) = b_\Sigma^{\,2} \qquad (5.59)$$

Likewise $\ln \Sigma_y$ has a normal probability density with

$$E(\ln \Sigma_y) = a_{\Sigma_y} \qquad (5.60)$$

$$Var(\ln \Sigma_y) = b_{\Sigma_y}^{\,2}$$

Hence, the difference $\ln Z = \ln \Sigma - \ln \Sigma_y$ has a normal probability density with

$$E(\ln Z) = a_\Sigma - a_{\Sigma_y} \qquad (5.61)$$

$$Var(\ln Z) = b_\Sigma^{\,2} + b_{\Sigma_y}^{\,2}$$

We also conclude that Z is log-normal random variable with

$$E(Z) = \exp[a_\Sigma - a_{\Sigma_y} + \tfrac{1}{2}(b_\Sigma^2 + b_{\Sigma_y}^2)] \tag{5.62}$$

$$Var(Z) = \exp[2(a_\Sigma - a_{\Sigma_y}) + b_\Sigma^2 + b_{\Sigma_y}^2][\exp(b_\Sigma^2 + b_{\Sigma_y}^2) - 1] \tag{5.63}$$

The reliability reads

$$R = \Phi\left[-\frac{a_\Sigma - a_{\Sigma_y}}{\sqrt{b_\Sigma^2 + b_{\Sigma_y}^2}}\right] \tag{5.64}$$

The central safety factor reads

$$n_1 = \frac{E(\Sigma_y)}{E(\Sigma)} = \frac{\exp(a_\Sigma + b_\Sigma^2/2)}{\exp(a_{\Sigma_y} + b_{\Sigma_y}^2/2)} \tag{5.65}$$

Consider an example. Let

$$\begin{aligned}
E(\Sigma) &= 60,000 \quad kPa \\
\sqrt{Var(\Sigma)} &= 20,000 \quad kPa \\
E(\Sigma_y) &= 100,000 \quad kPa \\
\sqrt{Var(\Sigma_y)} &= 10,000 \quad kPa
\end{aligned} \tag{5.66}$$

The central safety factor equals

$$n_1 = \frac{E(\Sigma_y)}{E(\Sigma)} = \frac{100,000}{60,000} = 1.67 \tag{5.67}$$

The reliability in this case turns out to be equal $R = 0.9495$.

5.6   The Characteristic Safety Factor and the Design Safety Factor

The characteristic safety factor reads

$$\gamma = \frac{\Sigma_{y,0.05}}{\Sigma_{0.95}} \tag{5.68}$$

where $\Sigma_{y,0.05}$ is 0.05 fractile of the probability distribution of yield stress, $\Sigma_{0.95}$ is 0.95 fractile of the probability distribution of stress.

When both random variables Σ and Σ_y are normal, then, according to Leporati (1979)

$$\gamma = \frac{E(\Sigma_y)(1 - 1.645 v_{\Sigma_y})}{E(\Sigma)(1 + 1.645 v_\Sigma)} = n_1 \frac{1 - 1.645 v_{\Sigma_y}}{1 + 1.645 v_\Sigma} \tag{5.69}$$

If, for example, both coefficients of variation are set at 0.05 then characteristic safety factor equals

$$\gamma = \frac{0.91775}{1.08225} n_1 = 0.848 n_1 \qquad (5.70)$$

For the coefficients of variation set at 0.1 we get

$$\gamma = 0.316 n_1 \qquad (5.71)$$

The design safety factor, according to Leporati (1979) equals

$$\gamma^* = \frac{\Sigma_{y,0.005}}{\Sigma_{0.95}} \qquad (5.72)$$

where $\Sigma_{y,0.005}$ is the 0.005 fractile of the yield stress. It equals

$$\gamma^* = n_1 \frac{1 - 2.576 v_{\Sigma_y}}{1 + 1.645 v_{\Sigma}} \qquad (5.73)$$

5.7   Asymptotic Analysis

Eq. (5.16) can be put in the following form:

$$P_f = \Phi(\beta) \qquad (5.74)$$

where β is referred to as a reliability index.

For $\beta > 5$, the probability of failure can be written via an asymptotic formula:

$$P_f \approx \frac{1}{\sqrt{2\pi}} \frac{\beta^2 - 1}{\beta^2} exp\left(-\frac{\beta^2}{2}\right) \qquad (5.75)$$

The reliability index itself is represented as follows, via Eq. (5.16)

$$\beta = \frac{E(\Sigma_y) - E(\Sigma)}{\sqrt{Var(\Sigma_y) + Var(\Sigma)}} \qquad (5.76)$$

Dividing both numerator and denominator by $E(\Sigma)$ we get

$$\beta = \frac{n_1 - 1}{\sqrt{v_{\Sigma}^2 + n_1^2 v_{\Sigma_y}^2}} \qquad (5.77)$$

From Eq. (5.77) we can find n_1 by solving the quadratic

$$\beta^2 v_{\Sigma}^2 + \beta^2 n_1^2 v_{\Sigma_y}^2 = (n_1 - 1)^2 \qquad (5.78)$$

$$n_1^2 (\beta^2 v_{\Sigma_y}^2 - 1) + 2 n_1 + \beta^2 v_{\Sigma}^2 - 1 = 0 \qquad (5.79)$$

We get

$$n_1 = \frac{1 + \sqrt{1 - \left(\beta^2 v_\Sigma^2 - 1\right) \times \left(\beta^2 v_{\Sigma_y}^2 - 1\right)}}{1 - \beta^2 v_{\Sigma_y}^2}$$

$$= \frac{1 + \sqrt{\beta^2 v_{\Sigma_y}^2 + \beta^2 v_\Sigma^2 - \beta^4 v_{\Sigma_y}^2 v_\Sigma^2}}{1 - \beta^2 v_{\Sigma_y}^2} \tag{5.80}$$

As is easily seen, not for all coefficients of variation v_Σ and v_{Σ_y} one can find the central factor of safety such that the demanded reliability index would be achieved. For example for

$$(\beta^2 v_\Sigma^2 - 1)(\beta^2 v_{\Sigma_y}^2 - 1) > 1 \tag{5.81}$$

n_1 gets complex values.

Solving the inequality (5.81) we get

$$\beta^2 (\beta^2 v_\Sigma^2 v_{\Sigma_y}^2 - v_{\Sigma_y}^2 - v_\Sigma^2) > 0 \tag{5.82}$$

This implies, that the following inequality must be met

$$\beta < \sqrt{\frac{1}{v_\Sigma^2} + \frac{1}{v_{\Sigma_y}^2}} \tag{5.83}$$

If the inequality (5.83) is violated, then the required reliability cannot be achieved by any factor of safety.

From Eq. (5.77) we observe that when n_1 tends to infinity, the reliability index tends to

$$\beta \rightarrow \frac{1}{v_\Sigma} \tag{5.84}$$

We can differentiate Eq (5.77) with respect to n_1:

$$\frac{\partial \beta}{\partial n_1} = \frac{v_\Sigma^2 + v_{\Sigma_y}^2}{(v_\Sigma^2 + n_1^2 v_{\Sigma_y}^2)^{3/2}} > 0 \tag{5.85}$$

This implies that when increasing n_1 from unity (when $\beta = 0$) to infinity, β varies from zero to the value $1/v_{\Sigma_y}$.

If the variation of the stress is zero, $v_\Sigma = 0$, and Eq. (5.80) yields

$$n_1 = \frac{1}{1 - \beta v_{\Sigma_y}} \tag{5.86}$$

When we have a zero variability of the yield stress $v_{\Sigma_y} = 0$, we get

$$n_1 = 1 + \beta v_\Sigma \tag{5.87}$$

5.8   Actual Stress and Yield Stress Are Correlated

"In most cases the correlation between the actual stress and the yield stresses is absent. Yet, when such a correlation exists, its expressing is ambiguous, and it is difficult to express it numerically", as Rzhanitzin (1981) notes in his textbook.

Positive correlation between Σ and Σ_y take place when the stronger elements take more load. Partially this takes place for statically indeterminate systems, in which greater strength is associated with greater stiffness, and hence with more loads. Safety margin

$$M = \Sigma_y - \Sigma \qquad (5.88)$$

has a variance

$$Var(M) = Var(\Sigma_y) - 2Cov(\Sigma_y, \Sigma) + Var(\Sigma) \qquad (5.89)$$

where $Cov(\Sigma_y, \Sigma)$ is the covariance between the actual stress and yield stress.

Then instead of Eq. (5.76) we get

$$\beta = \frac{E(\Sigma_y) - E(\Sigma)}{\sqrt{Var(\Sigma_y) - 2Cov(\Sigma_y, \Sigma) + Var(\Sigma)}} \qquad (5.90)$$

Hence, in terms of central safety factor, we have

$$\beta = \frac{n_1 - 1}{\sqrt{v_{\Sigma_y}{}^2 - 2n_1 v_{\Sigma\Sigma_y}{}^2 + n_1{}^2 v_\Sigma{}^2}} \qquad (5.91)$$

where $v_{\Sigma\Sigma_y}$ is the correlation coefficient

$$v_{\Sigma\Sigma_y} = \sqrt{\frac{Cov(\Sigma_y, \Sigma)}{E(\Sigma_y)E(\Sigma)}} \qquad (5.92)$$

One can express n_1 via β. Eq. (5.91) becomes

$$(1 - v_{\Sigma_y}{}^2 \beta^2)n_1{}^2 - 2(1 - \beta^2 v_{\Sigma\Sigma_y}{}^2)n_1 + (1 - \beta^2 v_\Sigma{}^2) = 0 \qquad (5.93)$$

yielding

$$n_1 = \frac{1 - \beta^2 v_{\Sigma\Sigma_y}{}^2 + \sqrt{\beta^2(v_{\Sigma_y}{}^2 - 2v_{\Sigma\Sigma_y}{}^2 + v_\Sigma{}^2) - \beta^4(v_\Sigma{}^2 v_{\Sigma_y}{}^2 - v_{\Sigma\Sigma_y}{}^4)}}{1 - \beta^2 v_{\Sigma_y}{}^2} \qquad (5.94)$$

When $v_{\Sigma\Sigma_y}$ vanishes Eq. (5.94) reduces to the case of uncorralated actual stress and yield stress. Another important case of correlation is studied by Grigoriu and Turkstra (1979).

5.9   Both Actual Stress and Yield Stress Follow the Pearson Probability Densities

The random variable is said to have a Pearson probability density, if it has a form

$$f_X(x) = Ax^a e^{-bx} \quad for \quad x > 0 \tag{5.95}$$

where

$$A = \frac{b^{a+1}}{\Gamma(a+1)} \tag{5.96}$$

The mean value $E(X)$ reads

$$E(X) = \frac{a+1}{b} \tag{5.97}$$

whereas the variance equals

$$Var(X) = \frac{a+1}{b^2} \tag{5.98}$$

The coefficient of variation equals

$$v_X = \frac{1}{\sqrt{a+1}} \tag{5.99}$$

It is interesting that the coefficient of variation does not depend upon b. It is seen that for small values of a we get very high variability, whereas for small variability, of the order of 0.1, greater values of a are needed ($a \approx 100$).

Let the actual stress have the Pearson density

$$f_\Sigma(\sigma) = A_1 \sigma^{\alpha-1} \exp\left(-\frac{\alpha\sigma}{E(\Sigma)}\right) \tag{5.100}$$

The yield stress also has the Pearson density

$$f_{\Sigma_y}(\sigma_y) = A_2 \sigma_y^{\beta-1} \exp\left(-\frac{\beta\sigma_y}{E(\Sigma_y)}\right) \tag{5.101}$$

where

$$A_1 = \frac{\alpha^\alpha}{\Gamma(\alpha)E^\alpha(\Sigma)}, \quad A_2 = \frac{\beta^\beta}{\Gamma(\beta)E^\beta(\Sigma)} \tag{5.102}$$

where

$$\alpha = \frac{1}{v_\Sigma^2}, \quad \beta = \frac{1}{v_{\Sigma_y}^2} \tag{5.103}$$

The probability of failure becomes

$$P_f = \frac{B_\delta(\beta,\alpha)}{B(\beta,\alpha)} = J_\delta(\alpha,\beta) \tag{5.104}$$

where

$$B(\beta,\alpha) = \frac{\Gamma(\alpha)\Gamma(\beta)}{\Gamma(\alpha+\beta)} \tag{5.105}$$

is the Euler function of the first kind, or beta-function, whereas

$$B_\delta(\beta,\alpha) = \int_0^\delta t^{\beta-1}(1-t)^{\alpha-1} dt \tag{5.106}$$

is the Euler function of the second kind, or incomplete beta-function. The quantity δ in Eq. (5.106) is defined as follows:

$$\delta = \frac{\beta E(\Sigma)}{\beta E(\Sigma) + \alpha E(\Sigma_y)} = \left(1 + \frac{v_{\Sigma_y}}{\Sigma_y^2} s_1\right)^{-1} \tag{5.107}$$

The evaluation for the function $J_\delta(\alpha,\beta)$ can be done by the numerical evaluation of integrals in Eqs. (5.106) and (5.107).

5.10 Conclusion: Reliability and Safety Factor Can Peacefully Coexist

At this junction we ask the most important question: Are the safety factor and the reliability concepts contradictory or they can coexist peacefully? The Fig. 5.1-5.12 depict the dependence of the reliability in Eq. (5.22) vs. the central safety factor s_1. As is seen, for various variabilities of the stress and the yield stress one can assign both reliability and the central safety factor. Indeed, Fig. 5.1 shows that of the coefficient of variability of the stress $\gamma_\Sigma = 0.04$ and the designer wants the safety factor to be set, say at 1.3, the demand that the reliability is above 0.92 results in the choice of materials with $\gamma_{\Sigma_y} \leq 0.15$. We immediately observe that both the central safety factor and the reliability requirements can be combined. Likewise, for $\gamma_\Sigma = 0.05$ (Fig. 5.2), the central safety factor $s_1 = 1.2$ is associated with reliabilites grater than or equal to 0.93. Thus, one can also impose the reliability constraint. Analogous features are characteristic to figures 5.3-12. This leads to the conclusion that these 2 concepts can coexist peacefully.

But this coexistence cannot be accomodated without some adjustments. Reliability concept provides the rigorous values of the factors, for otherwise, *i.e.* without the general context of reliability, the safety factor will remain as the factor of experience but still the factor of theoretical ignorance. Adopting reliability as the main concept, the allocated values of safety factor will naturally follow. In a sense probabilistic methods do not constitute a "revolution" but rather a natural "evolution."

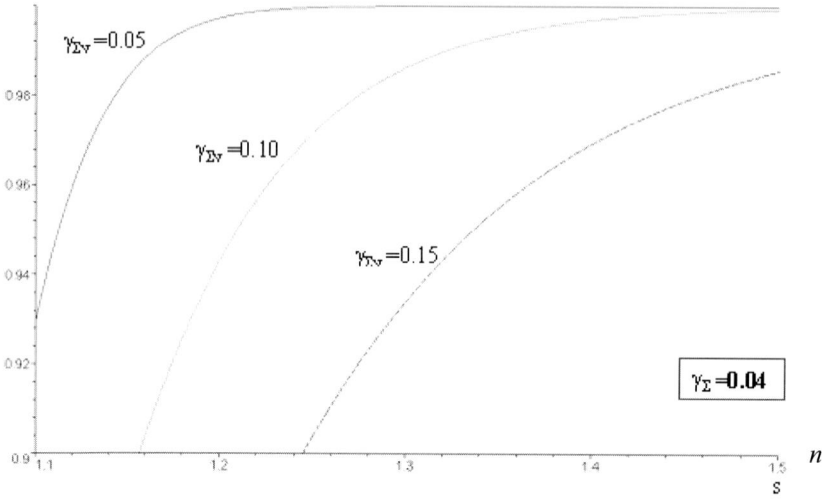

Fig. 5.1 Reliability R versus central safety factor n

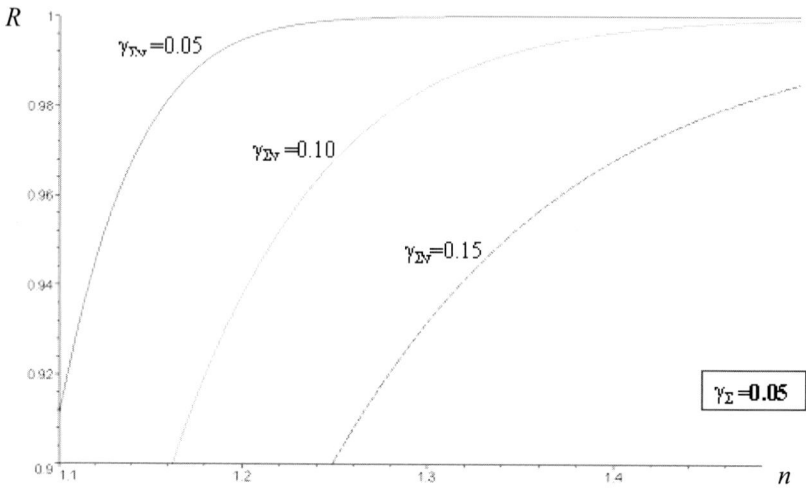

Fig. 5.2 Reliability R versus central safety factor n

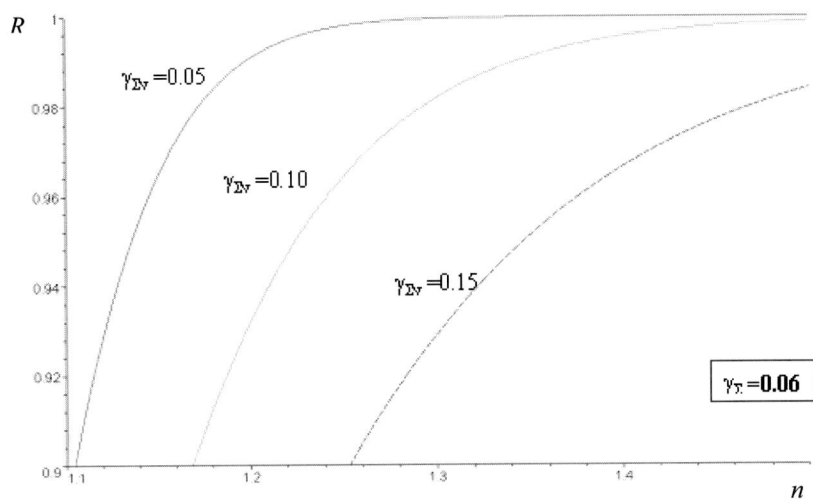

Fig. 5.3 Reliability R versus central safety factor n

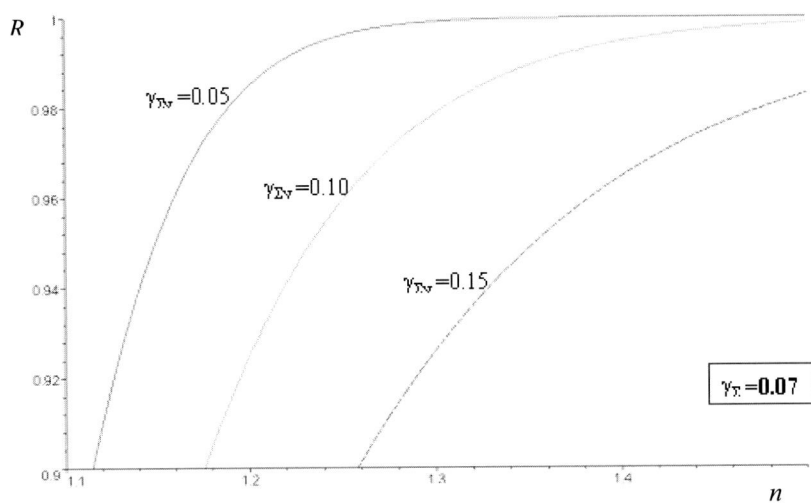

Fig. 5.4 Reliability R versus central safety factor n

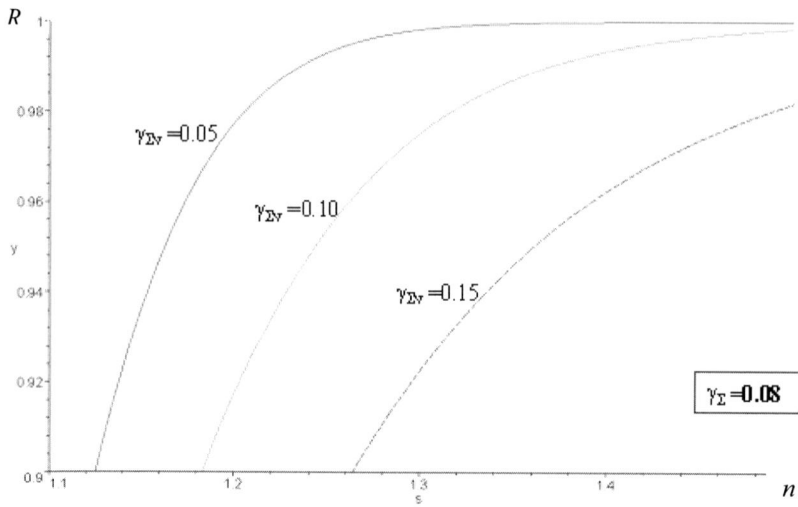

Fig. 5.5 Reliability R versus central safety factor n

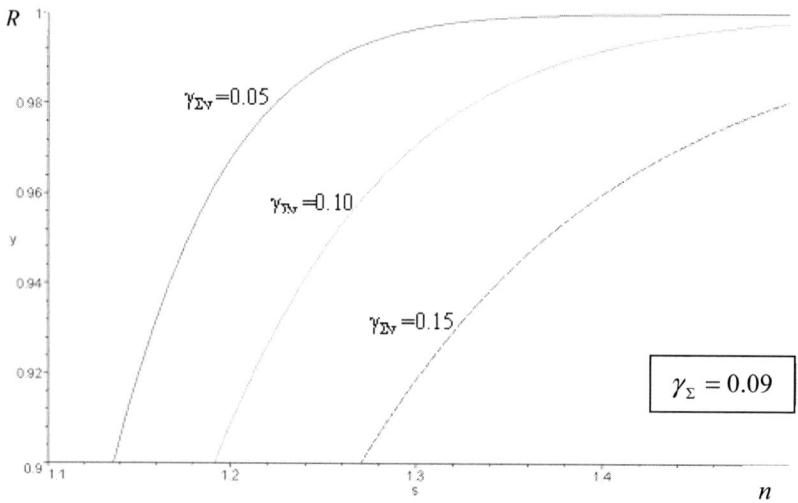

Fig. 5.6 Reliability R versus central safety factor n

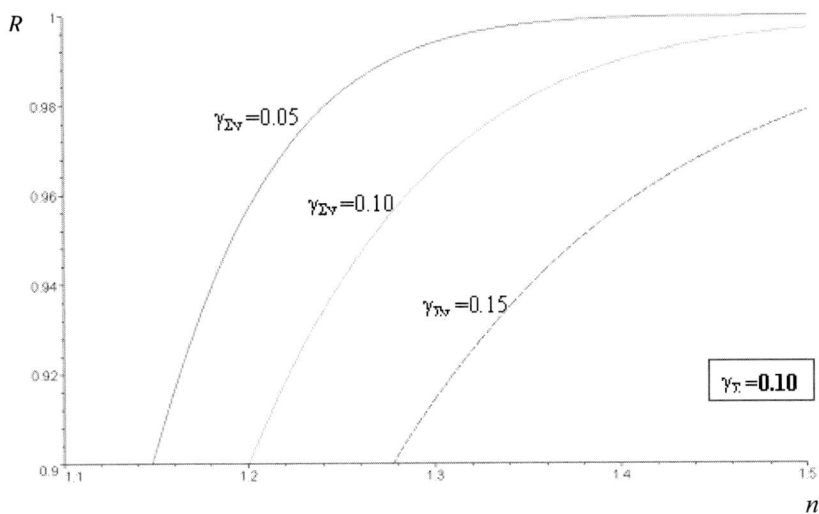

Fig. 5.7 Reliability R versus central safety factor n

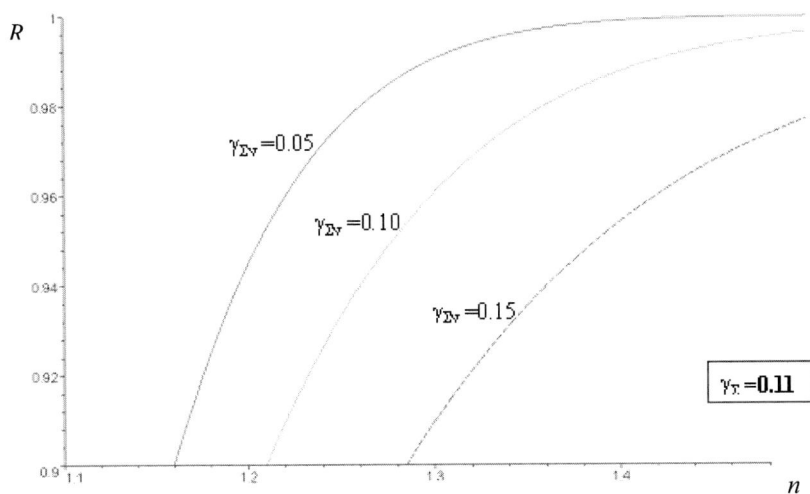

Fig. 5.8 Reliability R versus central safety factor n

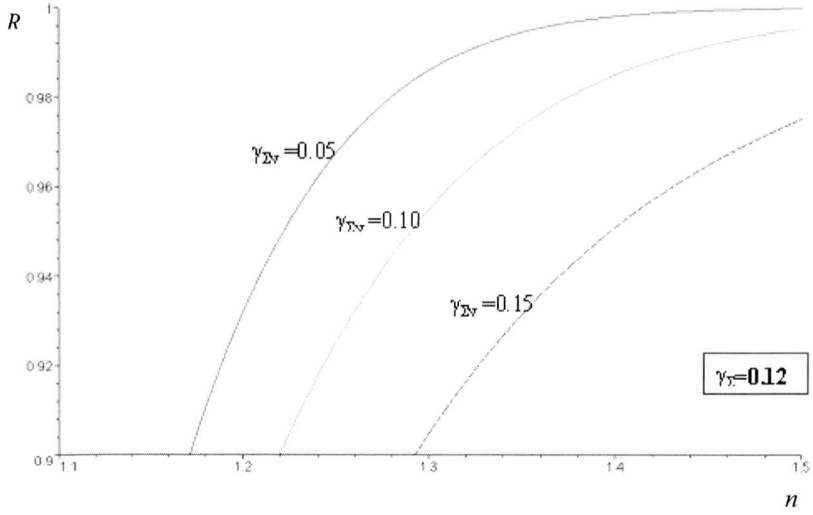

Fig. 5.9 Reliability *R* versus central safety factor *n*

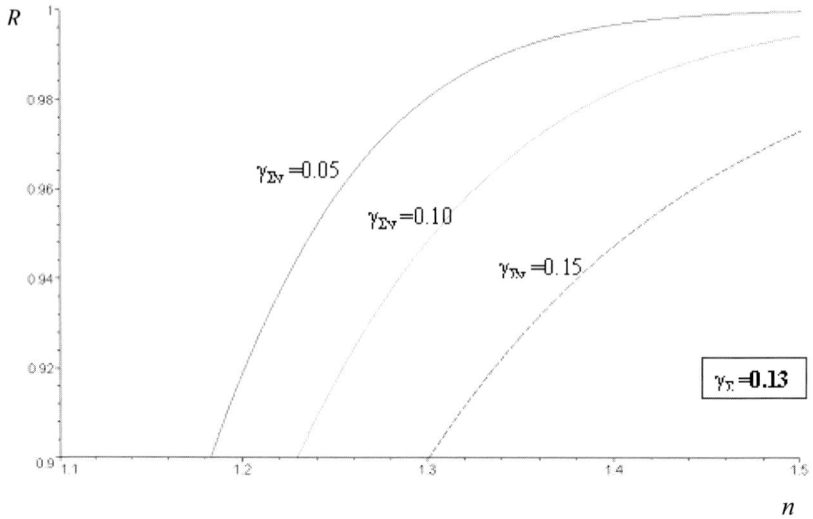

Fig. 5.10 Reliability *R* versus central safety factor *n*

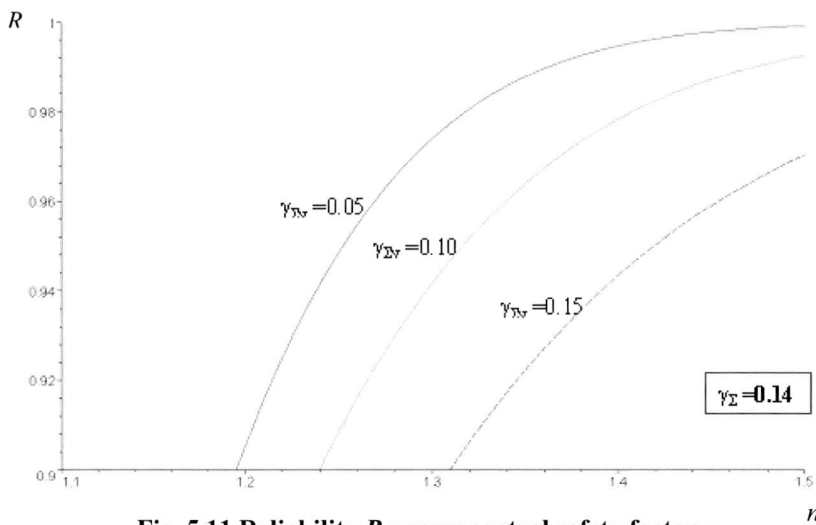

Fig. 5.11 Reliability R versus central safety factor n

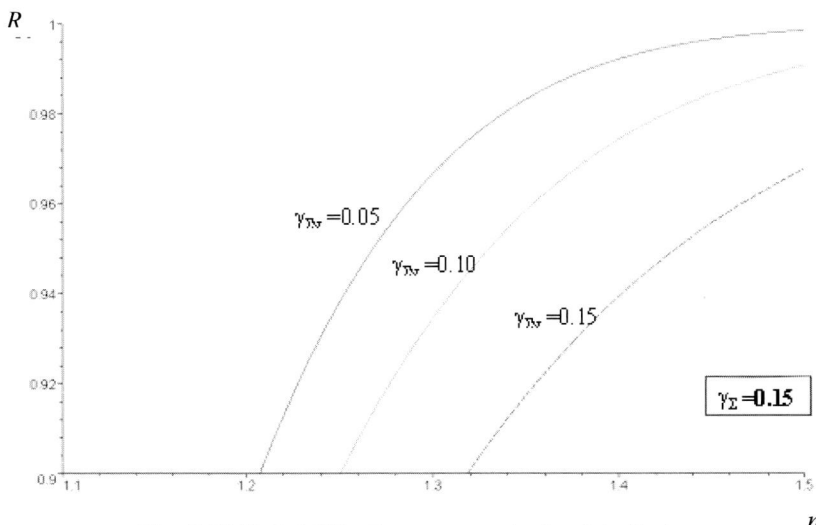

Fig. 5.12 Reliability R versus central safety factor n

Chapter 6
Non-Probabilistic Factor of Safety

"It is not implied that this use [of the theory of probability] is in itself sufficient to make a design more reliable or more economical, any more than that the avoidance of the probabilistic approach makes it safer."

A.M. Freudenthal (1972)

"In design situations, exact distributional forms are sometimes in doubt."

E.B. Haugen (1980)

"I also find that many designers have a 'gut feeling' that there is something not quite right with the statistics – the 'lies, damn lies, and statistics' school of thought – perhaps not without justification."

A.D.S. Carter (1997)

"One of the traditional methods to avoid failure in those cases there dispersions are expected is to do a "worst case design." The values of the various variables are taken at the $\pm 3\sigma$ of the dispersion range, and the structure is designed to survive these extreme values."

G. Maymon (2002)

"Good data on load and strength properties are often not available… in attempting to make account of variability, we are introducing assumptions that might be not tenable, e.g. by extrapolating the load and strength data to the very low probability tails of the assumed population distributions."

P.D.T. O'Connor (1985)

In this chapter we first illustrate some difficulties associated with probabilistic calculations of the reliability. It appears that these and other difficulties are either overlooked or pushed under the rug, as it were, by the probabilistic analysts. Then the alternative approach called 'convex modeling' is presented, and the safety factor's associated definition is suggested.

6.1   Introductory Comments

One can construct numerous simple examples that demonstrate sensitivity of the failure reliability to the tails of the probability distributions of the random values involved. It can be directly seen on an elementary example of evaluation of probabilistic characteristics. The example is due to Apostolakis and Kaplan (1981). Consider a cube with a side length X that is a random variable taking on two values $x_1 = 10^{-2}$ and $x_2 = 2$, with different probabilities:

$$Prob(X = 10^{-2}) = p_1 = 0.9999 \qquad (6.1)$$

$$Prob(X = 2) = p_2 = 0.0001$$

The mean value of the side length is

$$E(X) = x_1 p_1 + x_2 p_2 = 10^{-2} \times 0.9999 + 2 \times 0.0001 = 1.02 \times 10^{-2} \qquad (6.2)$$

The major contribution to this value $E(X)=1.02\times10^{-2}$ comes from the possible value $x_1=10^{-2}$, which has the highest probability. Suppose, now, that we are interested in the mean value of the cube's volume. The volume takes a value $x_1^3=(10^{-2})^3=10^{-6}$ with probability 0.9999; cube's volume takes on a value $x_2^3=8$ with probability 0.0001.The mean volume equals

$$E(X) = x_1 p_1 + x_2 p_2 = 10^{-2} \times 0.9999 + 2 \times 0.0001 = 1.02 \times 10^{-2} \qquad (6.3)$$

As is seen, the mean volume is mostly decided by the value $x_2=2$ which is taken with a very small probability $p_2 = 10^{-4}$, and not by the value $x_1 = 10^{-2}$ taken on with much higher probability, the factor p_1/p_2 being equals 9999!

Blekhman, Myshkis and Panovko (1983) write:

> "Significantly, the weakness of numerous works on stochastic models-sometimes ruling out any application-lies in the choice of statistical hypothesis, especially of assumptions regarding the probalistic features of the given accidental quantities and functions. These features are often regarded as fully known (like an assumption of a normal distribution will known parameters), or as amenable of determination. In real situations, it mostly turns out that the needed information is lacking."

Harris and Soms (1983) note:

> "…relatively small perturbations of tail of the strength distribution can make the failure probability far higher than may be desirable, particularly, when failure can be catastrophic."

Yao (1994) writes:

> "…it has been well known that the failure probability of engineering systems is highly sensitive to the tail portions of relevant distribution functions…It is not easy for me…to understand the necessity of computing the failure probability or reliability using various distributions, the tail portions of which remain to be difficult to ascertain."

Ang and Amin (1969) state:

> "In reality, of course, the random characteristics of [strength] R and [stress] S are unknown. Specifically, the form of the distribution functions and associated parameters are unknown; it is possible only to make predictions of what these are, or might be, using appropriate theoretical models. Such

theoretical predictions, however, are invariably imperfect and thus subject to errors…

At high-risk levels, e.g., $P_f \geq 10^{-3}$, differences in the shape of the distribution … would not sufficiently affect the calculated value of P_f; consequently, the choice of the distribution function for either variable…would not be too important.

At very low risk, say $P_f \leq 10^{-5}$, the calculated P_f will depend significantly on the distribution [of stress and of strength], so that the failure probability can be determined only with a knowledge of the correct distribution."

In the following section we will demonstrate the high sensitivity of the failure probability; the latter rather than mathematical expectation is of most importance to the application of design of structural components.

6.2   Sensitivity of Failure Probability

Consider an elastic bar compressed by an axial force. The uncertainty parameter is described by the non-vanishing eccentricities e_1 and e_2 for the force (Fig.6.1).

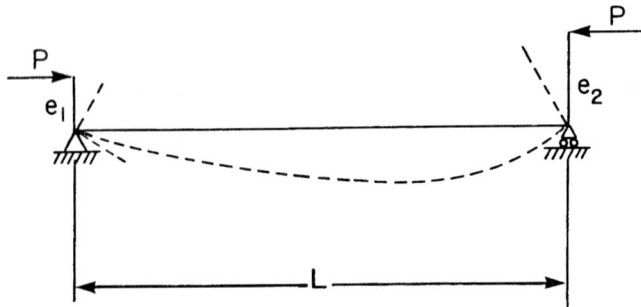

Fig. 6.1 Beam-column subjected to eccentric forces

The differential equation describing the deflection of the bar reads:

$$EI \frac{d^4 w}{dx^4} + P \frac{d^2 w}{dx^2} = 0, \qquad (0 \leq x \leq L) \tag{6.4}$$

where EI is the flexural stiffness, P the axial force, w the displacement and L the length of the bar. Denoting

$$k^2 = \frac{P}{EI} \tag{6.5}$$

The boundary conditions in terms of the bending moment are:

$$M_z(0) = Pe_1 \quad , \quad M_z(L) = Pe_2 \tag{6.6}$$

Compliance with the boundary conditions yields the final expression for $M_z(x)$:

$$M_z(x) = -\frac{P}{\sin kL}(e_2 - e_1 \cos kL)\sin kx + Pe_1 \cos kx \qquad (6.7)$$

The maximal bending moment M_Z* is

$$M_z*(e_1, e_2) = -\frac{P}{\sin kL}\sqrt{e_1^2 + e_2^2 - 2e_1 e_2 \cos kl}. \qquad (6.8)$$

This expression coincides with Eq. 44 in Pikovsky (1961). The problem of a bar in compression with two eccentricities was also studied by Young (1932) and Timoshenko and Gere (1963).

One can show that for the maximum bending moment to take place inside the bar, $0<x*<L$, the following conditions must be satisfied:

$$\cos kL < \frac{e_1}{e_2} < \frac{1}{\cos kL}, \qquad (6.9)$$

and

$$0 < P < \frac{\pi^2 EI}{4L^2} \quad , \quad 0 < kL < \frac{\pi}{2} \qquad (6.10)$$

It can be shown that the maximum bending moment occurs inside the bar and condition (6.9) is dispensed with, in the following range of load variation:

$$\frac{\pi^2 EI}{4L^2} < P < \frac{\pi^2 EI}{L^2} \qquad (6.11)$$

We assume now that the eccentricities constitute a random vector with a jointly exponential distribution and the following distribution function (Gumbel, 1960):

$$F_{E_1 E_2}(e_1, e_2) = \left[1 - \exp\left(\frac{e_1}{\beta}\right)\right]\left[1 - \exp\left(-\frac{e_2}{\gamma}\right)\right]\left[1 + \alpha \exp\left(-\frac{e_1}{\beta} - \frac{e_2}{\gamma}\right)\right], \qquad (6.12)$$

where e_1 and e_2 take on only positive values and $\beta = E(E_1)$, $Y = E(E_2)$, $E(\bullet)$ denoting mathematical expectation.

For the sake of simplicity, we will concentrate on the case presented by Eq. (6.6). We are interested in the reliability of the bar, defined as the probability of nonexceedence of a limiting value $m*$ by the random variable $M*$

$$R = \text{Prob}\left(M* = \frac{P}{\sin kL}\sqrt{E_1^2 + E_2^2 - 2E_1 E_2 \cos kL} \leq m*\right) \qquad (6.13)$$

The integration results in

$$F_{M*}(m*) = 1 - \frac{2\beta^3 + \alpha\beta^2\gamma - 5\beta^2\gamma - 2\alpha\beta\gamma^2 + 2\beta\gamma^2}{2\beta^3 - 7\beta^2\gamma + 7\beta\gamma^2 + 2\gamma^3}\exp\left(-\frac{m*\sin kL}{\beta P}\right)$$

$$-\frac{2\alpha\beta^2\gamma - 2\beta^2\gamma - \alpha\beta\gamma^2 + 5\beta\gamma^2 - 2\gamma^3}{2\beta^3 - 7\beta^2\gamma + 7\beta\gamma^2 + 2\gamma^3}\exp\left(-\frac{m*\sin kL}{\gamma P}\right)$$

$$-\frac{\alpha\beta\gamma(2\gamma-\beta)}{2\beta^3-7\beta^2\gamma+7\beta\gamma^2+2\gamma^3}\exp\left(-\frac{2m*\sin kL}{\gamma P}\right) \tag{6.14}$$

$$-\frac{\alpha\beta\gamma(\gamma-2\beta)}{2\beta^3-7\beta^2\gamma+7\beta\gamma^2+2\gamma^3}\exp\left(-\frac{2m*\sin kL}{\beta P}\right)$$

when following restriction holds

$$2\beta^3-7\beta^2\gamma+7\beta\gamma^2-2\gamma^3\neq \tag{6.15}$$

In the particular cases where instead of the inequality in Eq. (6.15), we have an equality, the expressions for the reliability read:

$$F_{M*}(m*)=1-\left(1+\frac{m*\sin kL}{\beta P}\right)\exp\left(-\frac{m*\sin kL}{\beta P}\right) \tag{6.16}$$

$$-\alpha\left[\left(\frac{m*\sin kL}{\beta P}-3\right)\exp\left(-\frac{m*\sin kL}{\beta P}\right)\right]+\left(\frac{2m*\sin kL}{\beta P}+3\right)\exp\left(-\frac{m*\sin kL}{\beta P}\right),\ for\ \beta=\gamma$$

and

$$F_{M*}(m*)=1-2\left(1-\frac{\alpha}{3}\right)\exp\left(-\frac{m*\sin kL}{\beta P}\right) \tag{6.17}$$

$$+\left(1+\frac{2\alpha m*\sin kL}{\beta P}\right)\exp\left(-\frac{2m*\sin kL}{\beta P}\right)+\frac{2\alpha}{3}\exp\left(-\frac{4m*\sin kL}{\beta P}\right),\ for\ \beta=2\gamma$$

Finally,

$$F_{M*}(m*)=1-2\left(1-\frac{\alpha}{3}\right)\exp\left(-\frac{m*\sin kL}{2\beta P}\right) \tag{6.18}$$

$$+\left(1+\frac{\alpha m*\sin kL}{\beta P}\right)\exp\left(-\frac{m*\sin kL}{\beta P}\right)+\frac{2\alpha}{3}\exp\left(-\frac{2m*\sin kL}{\beta P}\right),\ for\ \beta=\gamma/2$$

It can be shown (Gumbel, 1960) that the following restrictions should be met, $F_M*(m*)$ to serve as the distribution function:

$$1\leq\alpha\leq1$$
$$\alpha=4\rho$$

Also, k could be expressed as

$$k=\frac{\pi}{L}\sqrt{\frac{P}{P_{cl}}} \tag{6.19}$$

where P_{cl} is the classical buckling load of the simply supported bar $P_{cl}=\pi^2 EI/L^2$. The reliability equals

$$R = Prob\left(\Sigma \le \sigma_y\right) = Prob\left[\frac{M*c}{I} \le \sigma_y\right] = F_{M*}\left[\frac{\sigma_y I}{c}\right] \tag{6.20}$$

where Σ is the maximum stress, σ_y-yield stress assumed to be constant, I-moment of inertia and c- the distance between the centroidal line and the extreme fiber where the maximum stress occurs. Reliability of the structure is obtainable from Eq. (6.14)-(6.18) by replacing $m*$ by $\sigma_y I/c$.

Figures 6.2-6.4 depicts the reliability of the structure versus c.

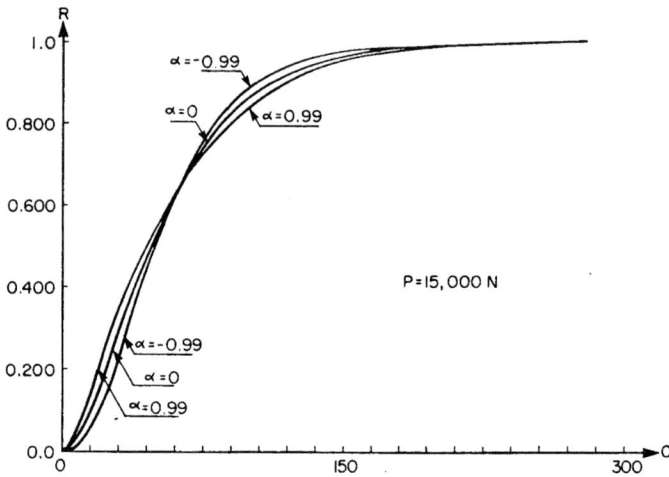

Fig. 6.2 Structural reliability versus radius

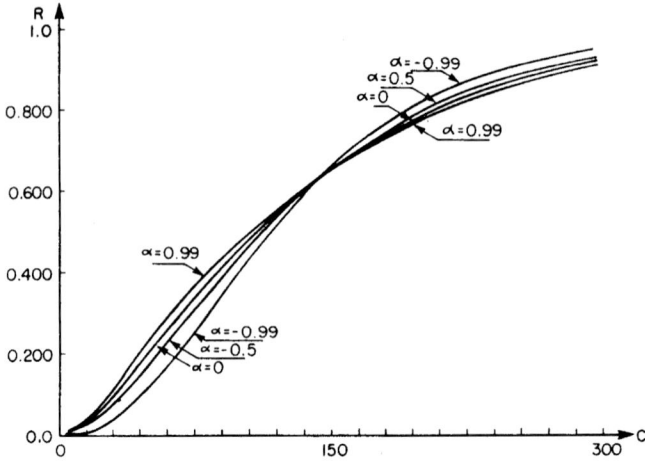

Fig. 6.3 Dependence of reliability upon parameters

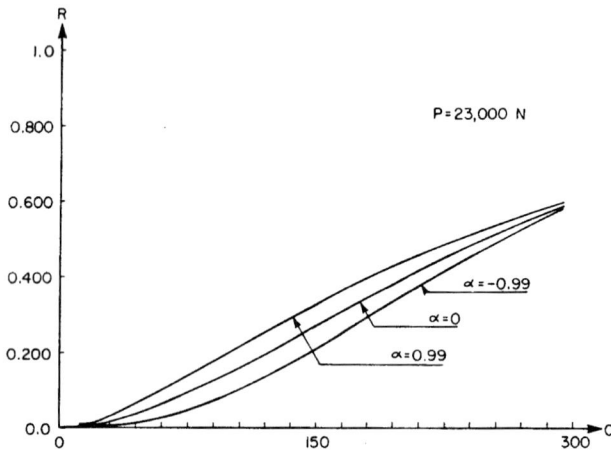

Fig. 6.4 Influence of the correlation coefficient

Fig 6.2 is associated with $P = 15$ kN, $P/P_{cl} = 0.569$. The following data is used in Fig. 6.6: $P = 20$ kN, $P/P_{cl} = 0.759$, whereas in Fig. 6.4 the data is fixed at $P = 23$ kN, $P/P_{cl} = 0.873$. In all three figures $\beta = 2$mm, $\gamma = 1.5$ mm, and $I/c = 1,333.3$ mm^3; $L = 1,000$ mm, $E = 200,000$ MPa. As we see, the increase in the applied loading results in reduced reliability of the structure. The coefficient α is varied in Figures 6.2-6.5, in the range $-0.99 < \alpha < 0.99$. Whereas data on β and γ may be reliable, information on the coefficient α could be insufficient, so that Figures 6.2-6.4 demonstrate the possible scatter in the reliability of the structure due to imprecise knowledge of parameter α. This is illustrated in Fig. 6.5, which addresses the design problem: Find the radius of the cross-section c so that the required codified reliability r, or codified probability of failure $P_{f*} = 1-r$ will be achieved.

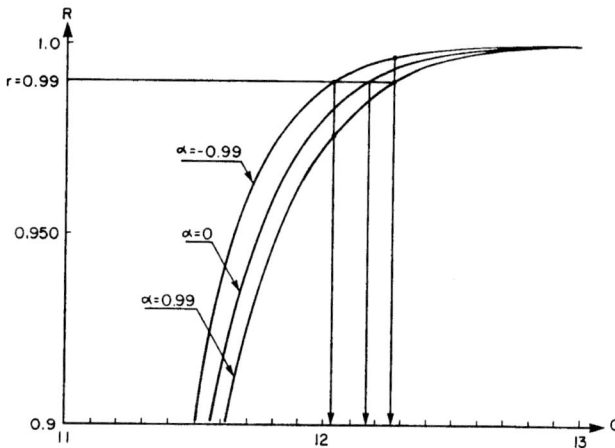

Fig. 6.5 Solution of the design problem

If we fix value of P_{f*} at 0.01, then, if the calculations are based on $\alpha = 0.99$, the design value of the radius is $c = 12.257$; now, if the true value of α is -0.99 then the actual probability of failure of the system at this radius is 3.74×10^{-3}, i.e. is lower than the codified probability of failure. This implies that we had a case of the "favorable" imprecision. However, if our calculations are based on value $\alpha = -0.99$, then the minimum required radius of the cross section should be $c = 12.029$. If however the true value of α is 0.99, then the chosen value of the radius corresponds to actual probability of failure 0.0234 instead of $P_{f*} = 0.01$. This corresponds to the underestimation of the probability of failure by more than twice.

The situation is more severe for highly reliable structures. To get more insight we define the underestimation factor as the ratio of the actual-to-codified probabilities of failure

$$\eta = \frac{P_f}{P_{f*}} \tag{6.21}$$

For $P_{f*} = 10^{-3}$, the underestimation factor is over three; for $P_{f*} = 10^{-4}$, $\eta = 3.47$; for $P_{f*} = 10^{-5}$, $\eta = 3.705$ and finally for $P_{f*} = 10^{-6}$ the underestimation factor reaches 3.82.

Thus, one would conclude that the system is acceptable for use, whereas the actual probability of failure is exceeding the codified one, and the system in fact is in a failed state, since the actual reliability is lower than the codified one.

Under these circumstances of the high sensitivity of probability of failure, the natural question arises on how the probabilistic analysis could have used for design purposes. To attempt to answer this question, we will visualize that the total cost T of production of the column is expressible as

$$T = \frac{q_1}{E(E_1) + E(E_2) + q_2} \tag{6.22}$$

where q_1 and q_2 are constants. Such a postulation maintains that more cost is associated with finer manufacturing, i.e. the one with less $E(E_j)$. Fig. 6.6 depicts the reliabilities of the columns associated with different mean values of the imperfections $E(E_j)$, while their sum $E(E_1) + E(E_2)$ is kept constant.

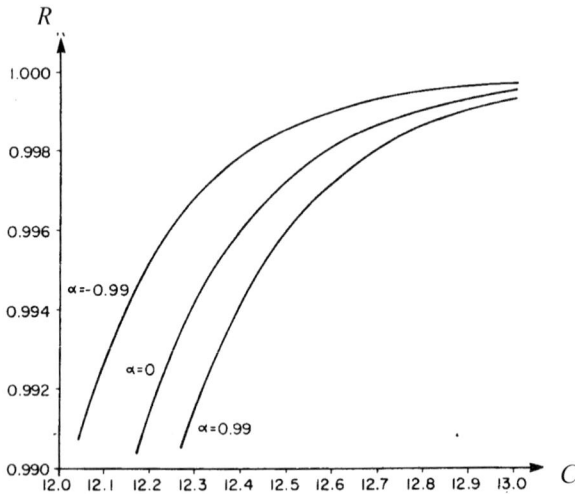

Fig. 6.6 How does the required high reliability influence the decision-making

Fig. 6.6 demonstrates that the maximum reliability is achieved for the equal mean imperfection parameters $E(E_1) = E(E_2)$. Thus, reliability studies could be utilized for comparative purposes; under other conditions being equal, one prefers the manufacturing process, which leads to higher reliability.

To degress, it is instructive to reproduce Jensen's (2000) comments:

"It mustn't break." This was a sentiment I heard expressed at a recent meeting, where a large corporation had invited one of its major customers to give a presentation and discuss the performance of the manufacturer's products. It was a product, where the competition is very, very fierce, and the revenue earned by the product to function continuously, twenty-four hours a day, every day of the year. The presenter summed up by saying: "I like your product, the price is right, just make sure it doesn't break."

There was some murmuring in the audience followed by some corporate hand waving quoting reliability figures, but the message indisputably expresses the down-to-earth sentiments of most customers of any technical product: our expectations are that it will work without failure, whenever called upon to do so, until we ourselves find the product obsolete and outdated and discard it for these reasons.

Of course, life is not simple, and no professional will make claims of product reliability being one hundred percent. But how close can we get? Well, this is what all our quality and reliability efforts revolve around."

As we see the reliability and the price we are ready to pay for it are intimately connected topics. As Rzhanitsyn (1978) suggests one ought to allocate an optimum reliability, corresponding to price minimization process. Examples of detailed calculations of this kind are given by Ang and De Leon (1997), Kanda and Ellingwood (1991), Kanda (1997) and other authors.

In the following section, we will discuss an alternative, non-probabilistic, method to deal with uncertainty in the same problem.

6.3   Remarks on Convex Modeling of Uncertainty

Number of linear problems has been considered under set-theoretical, convex modeling of uncertainty in structures, put forward in monograph (Ben-Haim and Elishakoff, 1990). Particularly, impact failure of bars (Ben-Haim and Elishakoff, 1989) and shells (Elishakoff and Ben-Haim 1990) was studied in detail, as was the response of a vehicle in uneven terrain (Ben-Haim and Elishakoff, 1989). By contrast, the only nonlinear problem considered in applied mechanics literature within the set-theoretical, convex modeling is buckling of shells with uncertain initial imperfections (Ben-Haim and Elishakoff, 1989). The latter paper studied the first-and second-order approximation for the nonlinear function, since the exact solution was unavailable. Present section contrasts the first-and second-order approximation discussed in detail by Ben-Haim and Elishakoff (1990). Here, we go beyond the latter monograph, presenting, apparently for the first time, the exact analysis for the model structure of the bar with two eccentricities.

Consider again an elastic bar under an axial compressive force, applied with eccentricities e_1 and e_2. The maximum bending moment M_z^* is given by Eq. (6.8). With e_1 and e_2 specified, the maximum value of the moment can be directly evaluated from Eq. (6.8). Assume now that the initial eccentricities are uncertain. In contrast to the previous section, we do not propose to model this uncertainty as randomness, under a probabilistic approach, but use an alternative, set-theoretical description, usually called "convexity modeling." The nominal values of the eccentricities are e_1^0 and e_2^0, respectively, and the deviations from these nominal values are denoted ζ_1 and ζ_2. We assume that these deviations vary within the ellipsoidal set:

$$Z(\alpha, \omega_1, \omega_2) = \left\{ (\zeta_1, \zeta_2) : \left(\frac{\zeta_1}{\omega_1}\right)^2 + \left(\frac{\zeta_2}{\omega_2}\right)^2 \leq \alpha^2 \right\} \qquad (6.23)$$

where ω_1 and ω_2 are semi-axes of the ellipsoid, and α is its size parameter. We are interested in finding the maximum $\mu(\alpha, \omega_1, \omega_2)$, with respect to the uncertainty in the eccentricity, of the maximum bending moment

$$\mu(\alpha, \omega_1, \omega_2) = \max M_z^* \left(e_1^0 + \zeta_1, \ e_2^0 + \zeta_2\right); \ \left\{\zeta_1, \zeta_2 \in Z(\alpha, \omega_1, \omega_2)\right\}, \qquad (6.24)$$

where $\mu(\alpha, \omega_1, \omega_2)$ is the bending moment of the weakest bar in the ensemble Z. The maximum bending moment for uncertain eccentricities ζ_1 and ζ_2 to the first order in ζ_1 and ζ_2 is

$$M_Z^*(e_1^0 + \zeta_1, e_2^0 + \zeta_2) = M_Z^*(e_1^0, e_2^0) + \frac{\partial M_Z^*}{\partial e_1}\bigg|_{e_i = e_i^0} \zeta_1 + \frac{\partial M_Z^*}{\partial e_2}\bigg|_{e_i = e_i^0} \zeta_2. \tag{6.25}$$

We will evaluate the maximum bending moment as ζ_1 and ζ_2 vary in an ellipsoidal set $Z(\alpha, \omega_1, \omega_2)$. For convenience, we define the vector γ as follows:

$$\gamma^T = \left\{ \frac{\partial M_Z^*}{\partial e_1}\bigg|_{e_i = e_i^0}, \quad \frac{\partial M_Z^*}{\partial e_2}\bigg|_{e_i = e_i^0} \right\} \tag{6.26}$$

where the superscript T denotes matrix transposition. Eq. (6.24), in view of Eqs. (6.25) and (6.26) becomes:

$$\mu(\alpha, \omega_1, \omega_2) = \max_{\zeta_1, \zeta_2 \in Z(\alpha, \omega_1, \omega_2)} \left[M_Z^*(e_1^0, e_2^0) + \phi^T \zeta \right], \tag{6.27}$$

where

$$\zeta^T = (\zeta_1, \zeta_2). \tag{6.28}$$

Define Ω as 2 x 2 diagonal matrix

$$\Omega = \begin{bmatrix} \dfrac{1}{\omega_1^2} & 0 \\ 0 & \dfrac{1}{\omega_2^2} \end{bmatrix} \tag{6.29}$$

Then Eq. (6.23) can be rewritten as

$$A(\alpha, \Omega) = \{\zeta : \zeta^T \Omega \zeta \le \alpha^2\} \tag{6.30}$$

Eq. (6.27) calls for finding the maximum of the linear functional $\gamma^T \zeta$ on the convex set $Z(\alpha, \omega_1, \omega_2)$. According to the well-known theorem (see e.g., Leunberger 1984; Arora, 1989) a linear functional, considered on the convex set Z, assumes the maximum on the set of extreme points of Z. The latter is the collection of vectors $\sigma = (\zeta_1, \zeta_2)$ in the following set:

$$C(\alpha, \Omega) = \{\sigma : \sigma^T \Omega \sigma = \alpha^2\}. \tag{6.31}$$

Thus the maximum bending moment becomes

$$\mu(\alpha, \Omega) = \max_{\sigma \in C(\alpha, \Omega)} [M_Z^*(e_1^0, e_2^0) + \gamma^T \sigma]. \tag{6.32}$$

To solve the problem, we use the method of Lagrange multipliers. For details of derivation, the reader should consult with the paper by Elishakoff, Ganashvili and Givoli (1991). Probabilistic analysis of the identical problem is given by Elishakoff and Nordstrand (1991).

For the maximum bending moment we arrive at the following expression

$$\mu(\alpha, \omega_1, \omega_2) = M_Z^*(e_1^0, e_2^0) + \gamma^T \sigma_1 = M_Z^*(e_1^0, e_2^0) + \alpha \sqrt{\gamma^T \Omega^{-1} \gamma}. \qquad (6.33)$$

In an analogous manner we arrive at the following expression for the minimum bending moment

$$\mu_{min}(\alpha, \omega_1, \omega_2) = M_Z^*(e_1^0, e_2^0) + \gamma^T \sigma_2 = M_Z^*(e_1^0, e_2^0) - \alpha \sqrt{\gamma^T \Omega^{-1} \gamma}. \qquad (6.34)$$

For the problem under consideration elements of vector φ can be found analytically:

$$\frac{\partial M_Z^*}{\partial e_1} = \frac{P \beta_1}{\sqrt{\beta_3} \sin kL}, \qquad (6.35)$$

$$\frac{\partial M_Z^*}{\partial e_2} = \frac{P \beta_2}{\sqrt{\beta_3} \sin kL}, \qquad (6.36)$$

where

$$\beta_1 = e_1 - e_2 \cos kL,$$

$$\beta_2 = e_2 - e_1 \cos kL, \qquad (6.37)$$

$$\beta_3 = e_1^2 + e_2^2 - 2 e_1 e_2 \cos kL,$$

Hence the maximum bending moment reads

$$M_{max}(\alpha, \omega_1, \omega_2) = M_Z^*(e_1^0, e_2^0)$$

$$+ \alpha \sqrt{\left[\omega_1 \frac{\partial M_Z(e_1, e_2)}{\partial e_1} \right]_{e_i = e_i^0}^2 + \left[\omega_2 \frac{\partial M_Z(e_1, e_2)}{\partial e_2} \right]_{e_i = e_i^0}^2}, \qquad (6.38)$$

whereas the minimum bending moment is

$$M_{max}(\alpha, \omega_1, \omega_2) = M_Z^*(e_1^0, e_2^0)$$

$$- \alpha \sqrt{\left[\omega_1 \frac{\partial M_Z(e_1, e_2)}{\partial e_1} \right]_{e_j = e_j^0}^2 + \left[\omega_2 \frac{\partial M_Z(e_1, e_2)}{\partial e_2} \right]_{e_j = e_j^0}^2}, \qquad (6.39)$$

The detailed second-order analysis can be found in the paper by Elishakoff, Ganashvili and Givoli (1991).

We will show below that the maximum bending moment is a convex function of its arguments ζ_1 and ζ_2. Indeed, according to a well-known theorem (Leunberger, 1984; Arora, 1989), a function of n variables defined on a convex set S is convex if and only if its Hessian matrix is positive semi-definite at all points in S. In our case the elements of the Hessian matrix γ_{ij} are

$$\gamma_{11} = \frac{e_2^2 (1 + \cos kL)^2}{\left[e_1^2 + e_2^2 - 2 e_1 e_2 \cos kL \right]^{3/2}}, \qquad (6.40)$$

$$\gamma_{12} = -(e_1^2 + e_2^2 - 2 e_1 e_2 \cos kL)^{-1/2} \left[\cos kL + \frac{(e_1 - e_2 \cos kL)(e_2 - e_1 \cos kL)}{e_1^2 + e_2^2 - 2 e_1 e_2 \cos kL} \right], \qquad (6.41)$$

$$\gamma_{22} = \frac{e_1^2 (1 + \cos kL)^2}{\left[e_1^2 + e_2^2 - 2 e_1 e_2 \cos kL \right]^{3/2}}, \qquad (6.42)$$

Direct calculation yields

$$det[\,\Gamma\,] = \gamma_{11} \gamma_{22} - \gamma_{12}^2 \equiv 0. \qquad (6.43)$$

Also

$$\gamma_{11} > 0, \ \gamma_{22} > 0. \qquad (6.44)$$

Eqs. (6.43) and (6.44) imply that the function M_Z^* is convex. Therefore, we can apply the following theorem (Leunberger, 1984; Arora, 1989); "Let f be a convex function defined on a bounded closed convex set S. If f has a maximum over S, this maximum is achieved at an extreme point of S."

Such being the case, we deduct that the maximum moment is achieved on the ellipse

$$\left(\frac{\zeta_1}{\omega_1} \right)^2 + \left(\frac{\zeta_2}{\omega_2} \right)^2 = \alpha^2 \qquad (6.45)$$

We express ζ_2 from the latter equation as

$$\zeta_2 = \pm \omega_2 \sqrt{\alpha^2 - \frac{\zeta_1^2}{\omega_1^2}} \qquad (6.46)$$

and substitute in Eq. (6.24) to yield

$$M_Z^*(e_1^0 + \zeta_1, e_2^0 + \zeta_2) = M_Z^* \left(e_1^0 + \zeta_1, e_2^0 \pm \omega_2 \sqrt{\alpha^2 - \zeta_1^2 / \omega_2^2} \right) \qquad (6.47)$$

Now we seek the maximum of M_Z^* with respect to ζ_1 alone. For the maximum moment we demand that

$$\frac{\partial M_Z^*}{\partial \zeta_1} = 0. \qquad (6.48)$$

This equation defines the ζ_1^* at which M_Z^* assumes maximum; then Eq. (6.43) determines the value of ζ_2^*. Substituting ζ_1^* and ζ_2^* into Eq. (6.24), we obtain the maximum value of the bending moment.

Fig. 6.7 shows the variation of the bending moment over the region

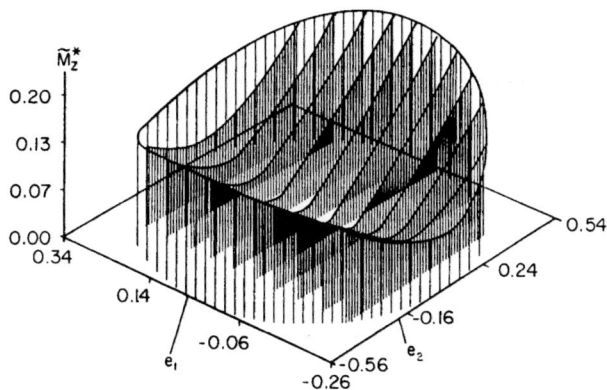

Fig. 6.7 Variation of the bending moment over the uncertainty ellipse

$$\left(\frac{\varsigma_1}{\omega_1}\right)^2 + \left(\frac{\varsigma_2}{\omega_2}\right)^2 \le \alpha^2 \qquad (6.49)$$

for $\omega_1 = 0.3$, $\omega_2 = 0.5$, at the non-dimensional load level

$$v = \frac{P}{P_E}, \quad P_E = \frac{\pi^2 EI}{L^2} \qquad (6.50)$$

equal $v = 0.3$. Fig. 6.8 is associated with $\omega_1 = 0.3$, $\omega_2 = 0.6$ and $v = 0.3$, whereas Fig. 6.9 illustrates the variation of M_Z^* for the values $\omega_1 = 0.3$, $\omega_2 = 0.6$ and $v = 0.5$. As we see in all these three instances the maximum value is achieved at the boundary point of the ellipse.

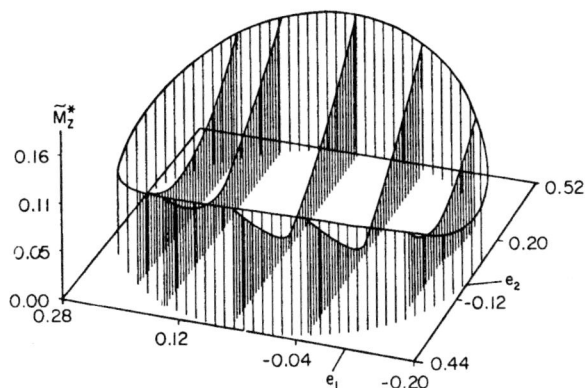

Fig. 6.8 Moment uncertainty as a function of uncertainty in eccentricities

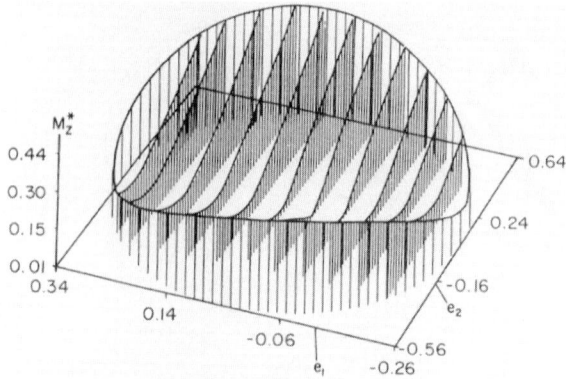

Fig. 6.9 Moment uncertainty as a function of uncertainty in eccentricities

The variation of the non-dimensional maximal moment

$$\overline{M}_z^* = \frac{M_z^*}{P_e c} \tag{6.51}$$

versus the ellipse's size α, where c is the radius of inertia of the bar's cross section is given in Fig. 6.10. Moreover, $\omega_1 = 0.001$, $\omega_2 = 0.02$, $e_1^0 = 0.04$, $e_2^0 = 0.04$. The non-dimensional load is fixed at $v = 0.3$.

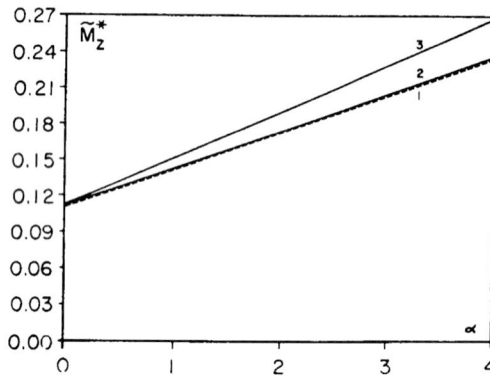

Fig. 6.10 Comparison of first order and second-order solutions with exact solution

Broken curves 1 are associated with the first-order analysis, curves marked 2 with the second order analysis and curves marked 3 with the exact results. The maximum moment increases with the size α of the uncertainty ellipse. For moderate values of α the agreement between the first-order, second-order and the exact analysis is excellent. It turns out that with the increase of the non-dimensional applied load, the percentage-wise disagreement between the first order and the second-order analyses decreases considerably. On the other hand, for the larger non-dimensional applied loads the difference between the low-order approximations and the exact analysis becomes wider.

We conclude, therefore, that the lower-order approximations yield acceptable results for small uncertainties and smaller loads. This modifies the conclusion, based on the comparison of the first-order and the second-order analyses, drawn in by Ben-Haim and Elishakoff (1990) that the first-order approximation was acceptable for small uncertainties and greater loads. It is remarkable that the similarity of the first-order and second-order approximations that occurs for the elastic bar does not generally suggest that these approximations are in good agreement with the exact solution.

6.4   "Worst-Case" Probabilistic Safety Factor

The following question arises: It is possible to perform worst-case analysis within the probabilistic context? At first glance this question may appear to be not valid. Indeed, if the random loads involved may take values from -∞ to +∞, naturally there is no "worst" value of the load. Some engineers suggest, therefore, to utilize some characteristic "worst" case analysis.

However, in practical applications, precisely, such an approach is often adopted. Bolotin (1968) notes:

> "In engineering codes as a design load some "maximum" (i.e. corresponding to certain small probability of realization) value is taken, whereas as a design resistance of the material some lower value, determined from technical conditions is utilized. Choice of these values, as well as of safety factor, is in some degree arbitrary: Various values of safety factor could correspond to some value of reliability."

Consider the following example treated by Maymon (1998). We will study the probabilistic behavior of an elastic bar of the rectangular cross section with width B and depth H. The bar is under the tensile force F. The quantities B, H and F are all normally distributed with

$$\sigma_B = 0.01 \, E(B)$$
$$\sigma_H = 0.01 \, E(H) \qquad\qquad (6.52)$$
$$\sigma_F = 0.01 \, E(F)$$

The normal stress in the bar

$$\Sigma = F / BH \qquad\qquad (6.53)$$

is naturally, not distributed normally. Yet, due to small standard deviations involved, it can be approximately treated as normally distributed random variable. If $E(\Sigma)=2083$ kg/cm^2 and $\sigma_\Sigma=104$ kg/cm^2, then we can conditionally agree to define the maximum tensile stress as the mean value of the stress *plus* the three times the standard deviation of it:

$$\sigma_{max} = 2083 + 3 \times 104 = 2395 \text{ kg/cm}^2 \qquad\qquad (6.54)$$

The failure of the bar is identified with the exceedance by the tensile stress the yield stress. The latter, Σ_y, is also assumed to be normally distributed, with

$$E(\Sigma_y) = 2500 \text{ kg/cm}^2$$
$$\sigma_{\Sigma_y} = 75 \text{ kg/cm}^2 \tag{6.55}$$

We introduce the conditionally "minimum" yield stress as the mean value *minus* three times its standard deviation:

$$\sigma_{min} = 2500 - 3 \times 75 = 2275 \text{ kg/cm}^2 \tag{6.56}$$

According to Maymon (1998), "three types of designers who would treat this problem differently are identified."

(a) The "nominal designer" evaluates the safety factor of the bar from nominal, mean values

$$n_{nominal} = E(\Sigma_y)/E(\Sigma) = 2500/2083 = 1.2 \tag{6.57}$$

which coincides with the central safety factor. If this value is not less than the required factor of safety specified for the specific project, the design is accepted. However, the design neglects the scatter expected during the production phase in variables B, H, F and Σ_y, which constitute as it was mentioned above random variables.

(b) The second type of the designer is identified by Maymon (1998) as the "3σ worst case designer." Such a designer calculates the safety factor as follows

$$n_{3\sigma} = \sigma_{y,min}/\sigma_{max} = 2275/2395 = 0.95 \tag{6.58}$$

which is not acceptable, since it is less than unity. One can make a correction in design. Increasing of the cross-section by 26% obtainable through forming the ratio 1.2/0.95=1.26 increases the weight by 26% and, consequently, the cost of the product. Another possibility is to decrease the allowed tolerances in B and H. One can show that in order to obtain the unity safety factor, which is less than required value of 1.2, the tolerances on the cross-section should be decreased from 1% to 0.04%. Even this change, however, may contribute to the tremendous increase in the product cost.

(c) The third type of designers is utilizing the reliability concept directly. Assume that the required value of reliability is 99.9%; then the 'probabilistic designer' (in Maymon's terminology) uses the probabilistic analysis. Probabilistic analysis demonstrates that

$$Prob(\Sigma < \Sigma_y) = 99.943\% \tag{6.59}$$

Since this value is higher than required 99.9% value, the designer is allowed to somewhat decrease the cross-section or slightly increase the distance on the cross-section area.

Let us prove the validity of Eq. (6.59) (Maymon, 2002b). The failure function g delimiting the safe and unsafe regions of operation, reads

$$g(\Sigma, \Sigma_y) = \Sigma_y - \Sigma = 0 \tag{6.60}$$

We transform the discussion into the space of standardized variables

$$Z_1 = \frac{\Sigma_y - E(\Sigma_y)}{\sigma_{\Sigma_y}}$$

$$Z_2 = \frac{\Sigma - E(\Sigma)}{\sigma_\Sigma} \tag{6.61}$$

Substituting Eq (6.61) into Eq (6.60) we get

$$G(z_1, z_2) = \sigma_{\Sigma_y} Z_1 - \sigma_\Sigma Z_2 + E(\Sigma_y) - E(\Sigma) = 0 \tag{6.62}$$

Hereinafter, we use the Lagrange multiplier method. Since the minimum distance from the origin to the failure boundary given in Eq (6.60) is given by expression

$$d = \sqrt{z_1^2 + z_2^2} , \tag{6.63}$$

we construct the following Lagrangean:

$$D = Z_1^2 + Z_2^2 - \lambda \left| \sigma_{\Sigma_y} Z_1 - \sigma_\Sigma Z_2 + E(\Sigma_y) - E(\Sigma) \right| \tag{6.64}$$

where λ is the Lagrange multiplier. Eq (6.62) in addition to the following two equations

$$\frac{\partial D}{\partial Z_1} = 2Z_1 - \lambda \sigma_{\Sigma_y} = 0 \tag{6.65}$$

$$\frac{\partial D}{\partial Z_2} = 2Z_2 + \lambda \sigma_\Sigma = 0 \tag{6.66}$$

provide three equations for three values: Z_1, Z_2 and the value of λ. We get

$$Z_1 = -1.9022566$$
$$Z_2 = 2.63779575 \tag{6.67}$$
$$\lambda_3 = -0.05072684$$

The reliability index becomes

$$\beta = \sqrt{Z_1^2 + Z_2^2} = 3.25216 \tag{6.68}$$

which corresponds to probability of failure 0.0572719 by using

$$P_f = \Phi(-\beta) \tag{6.69}$$

or a reliability of 99.943% as it was required to demonstrate.

Maymon (1998) concludes:

"Although this example is a simplification of real-life application, it shows that contradictory measures can be taken when using each approach. It also indicates that the probabilistic approach may be more realistic and may result in less conservative design. This approach is valid if project requirements are formulated by probabilistic methods. Although this is not the situation at present, it is believed that future requirements will contain more and more of this approach."

It should be stressed, that it is not necessary the worst case design to be associated with the mean plus or minus three standard deviations. Consider the following example due to Hart (1982).

A truss member of area a is subjected to an axial load P. Since P is given as a normal random variable, so is the resulting stress, denoted as Σ: $\Sigma = P/a$. Let Σ_y be also a normal random variable, independent of P. The probabilistic characteristics are

$$E(P) = 200 \ kips, \quad \sigma_P = 40 \ kips$$
$$E(\Sigma_y) = 47.9 \ ksi, \quad \sigma_{\Sigma_y} = 3.3 \ ksi \tag{6.70}$$

The member is designed in such a manner the central safety factor to constitute 1.2. We need to calculate the required truss member area. The central safety factor is

$$n = 1.2 = \frac{E(\Sigma_y)}{E(P)/a} \tag{6.71}$$

From here we find

$$a = 1.2 \frac{E(P)}{E(\Sigma_y)} = \frac{1.2 \times 200}{47.9} = 5.01 \ in^2 \tag{6.72}$$

Deterministic design load is defined as follows

$$P_{design} = E(P) + \alpha_P \sigma_P \tag{6.73}$$

when α_P is a constant such that the probability that the load is greater than P_{design} is 0.10. We observe that the probability that the load does not exceed that value P_{design} equals 0.9. Hence

$$Prob(P \le P_{design}) = 0.9 = \int_{-\infty}^{P_{design}} \frac{1}{40\sqrt{2\pi}} \exp\left[-\frac{1}{2}\left(\frac{t-200}{40}\right)^2\right] dt \tag{6.74}$$

Standardized normal-distribution table shows that

$$P_{design} = 251.3 \ kips \tag{6.75}$$

Hence,

$$P_{design} = E(P) + \alpha_P \sigma_P = 200 + \alpha_P \times 40 = 251.3 \tag{6.76}$$

Thus

$$\alpha_P = 1.28 \tag{6.77}$$

We represent P_{design} as follows:

$$P_{design} = \psi_{Load} E(P) \tag{6.78}$$

where the coefficient ψ_{Load} is called load factor. Substitution yields

$$P_{design} = 251.3 = \psi_{Load} E(P) = \psi_L \times 200 \tag{6.79}$$

resulting in

$$\psi_{Load} = 1.26 \tag{6.80}$$

We also define the deterministic design yield stress level to be

$$\sigma_{y,design} = E(\Sigma_y) - \alpha_{\Sigma_y}\sigma_{\Sigma_y} \qquad (6.81)$$

where the coefficient α_{Σ_y} is defined so that the probability that the yield stress is less than the designed value, equals 0.1. This leads to

$$0.1 = Prob(\Sigma_y \le \sigma_{y,design}) = \int_{-\infty}^{\sigma_{y,design}} \frac{1}{3.3\sqrt{2\pi}} exp\left[-\left(\frac{t-47.9}{3.3}\right)^2\right] dt \qquad (6.82)$$

Using again the standardized normal distribution table we find

$$\sigma_{y,design} = 43.7 \; ksi \qquad (6.83)$$

Hence

$$\sigma_{y,design} = E(\Sigma_y) - \alpha_{\Sigma_y}\sigma_{\Sigma_y} = 47.9 - \alpha_{\Sigma_y} \times 3.3 = 43.7 \qquad (6.84)$$

leading to

$$\alpha_{\Sigma_y} = 1.28 \qquad (6.85)$$

This allows one to introduce the capacity reduction factor $\psi_{capacity}$, if we define

$$\sigma_{y,design} = \psi_{capacity} E(\Sigma_y) \qquad (6.86)$$

leading to, numerically

$$43.7 = \psi_{capacit} \times 47.9 \qquad (6.87)$$

Hence,

$$\psi_{capacity} = 0.91 \qquad (6.88)$$

Now we can determine the required truss member area such that

$$\sigma_{design} = \frac{P_{design}}{a} \qquad (6.89)$$

Thus,

$$a = \frac{P_{design}}{\sigma_{design}} = \frac{251.3}{43.7} = 5.75 \; in^2 \qquad (6.90)$$

Now we can answer the question: "What is the probability of failure if the latter is defined to be a load-induced stress equal to or greater then a material yield stress when the area of the truss member is determined" (Hart, 2002). To answer this question we introduce the failure boundary function

$$g = \Sigma_y - \frac{P}{a} \qquad (6.91)$$

Now,

$$E(g) = E(\Sigma_y) - \frac{E(P)}{a}$$

$$\sigma_g^2 = \sigma_{\Sigma_y}^2 + \frac{\sigma_P^2}{a^2}$$

(6.92)

The probability of failure is, therefore,

$$P_f = \int_{-\infty}^{\sigma} \frac{1}{\sigma_g \sqrt{2\pi}} exp\left[-\frac{1}{2}\left(\frac{t - E(g)}{\sigma_g}\right)^2 \right] dg$$

(6.93)

Calculation yields

$$P_f = 0.178$$

(6.94)

Likewise if

$$a = 5.75 \ in^2, \ E(g) = 13.12, \ \sigma_g = 7.70$$

(6.95)

then

$$P_f = 0.044$$

(6.96)

Hart (2002) summarizes the evaluation of this example as follows:

> "Note that the two alternative deterministic design approaches result in two different member areas and hence two different probabilities of failure. The second approach directly incorporates the uncertainty in the loading and strength of the system."

6.5 Which Concept Is More Feasible: Non-Probabilistic Reliability or Non-Probabilistic Safety Factor?

In recent decade the non-probabilistic reliability concept was advocated by Ben-Haim (1994). He writes:

> "Reliability has a plain lexical meaning, which the engineers have modified and absorbed into their technical jargon. Lexically, that which is "reliable" can be depended upon confidently. Applying this to machines or system, they are "reliable" (still avoiding technical jargon) if one is confident that they will perform their specified tasks as intended. In current technical jargon, a system is reliable if the probability of failure is acceptably low. This is legitimate extension of the lexical meaning, since "failure" would imply behavior beyond the domain of specified tasks. The particular innovation which makes the development of modern engineering reliability is the insight that probability – a mathematical theory – can be utilized to quantify the qualitative lexical concept of reliability."

Ben-Haim (1994) further advocates:

> "We do not detract from the importance of the probabilistic concept of reliability by suggesting that probability is not the only starting point for quantifying the intuitive idea of reliability. Probabilistic reliability emphasized the *probability* of acceptable behavior. Non-probabilistic reliability, as developed here [Ben-Haim 1994], stresses the *range* of acceptable behavior. Probabilistically, a system is

reliable if the probability of unacceptable behavior is sufficiently low. In the non-probabilistic formulation of reliability, a system is reliable if the range of performance fluctuations is acceptably small."

In another study (1997) the following statement is made:

"The current standard theory of reliability is based on probability: The reliability of a system is measured by the probability of non-failure...In this paper we will describe a different formulation. We measure the reliability of a system by the amount of uncertainty consistent with no-failure. A reliable system will perform satisfactorily in the presence of great uncertainty. Such a system is robust with respect to uncertainty, and hence the name robust reliability."

In the discussion of the above 1994 paper the present writer (1995) noted:

"It appears...that the non-probabilistic concept of reliability is not necessary....Engineers are accustomed to highly reliable structures, with reliabilities of order $1-10^{-7}$, which has a frequency interpretation if the ensamble of produced structures is sufficiently large. A natural question arises: will non-probabilistic reliability take on comparable values?...Indeed, there exists a universally accepted probabilistic definition of reliability. The alternative, non-probabilistic definition of reliability is not formulated...What one may need, however, is the non-probabilistic concept of safety factor."

The question arises: How to introduce the non-probabilistic safety factor? To reply to this inquiry, consider a system subjected to combination of loads which belongs to some convex set. Assume, that the mathematical model is convex too. By utilizing convex analysis we determine the interval of variation of the stress Σ:

$$\Sigma = \left[\underline{\sigma}, \overline{\sigma}\right] \qquad (6.97)$$

where $\underline{\sigma}$ is the minimum stress that the system may experience, when loads vary in the convex set, whereas $\overline{\sigma}$ is the maximum stress that the system may assume. Let the yield stress Σ_y be also an interval uncertain variable

$$\Sigma_y = \left[\underline{\sigma}_y, \overline{\sigma}_y\right] \qquad (6.98)$$

Then the safety factor n can be defined in two different ways. One of the possible definitions is as a ratio of two interval variables

$$N = \Sigma_y / \Sigma \qquad (6.99)$$

Thus, the safety factor turns out to be an interval variable

$$N = \left[\underline{n}, \overline{n}\right] \qquad (6.100)$$

where $\underline{n}$ is the minimum value N may assumes while $\overline{n}$ is the maximum value. Naturally,

$$\underline{n} = \underline{\sigma}_y / \overline{\sigma}, \qquad \overline{n} = \overline{\sigma}_y / \underline{\sigma} \qquad (6.101)$$

The definition was proposed by Elishakoff (1995). Another definition is an analogue probabilistic central safety factor, and could be defined as follows

$$n = \frac{(\overline{\sigma}_y - \underline{\sigma}_y)/2}{(\overline{\sigma} - \underline{\sigma})/2} \qquad (6.102)$$

where the numerator represents the mid-value of the yield stress interval, while the denominator constitutes a mid-value of the stress interval.

Still, a nagging question may present: Can one define the non-probabilistic reliability? It is noted that the papers (Ben-Haim 1994, 1997) did not introduce the formal definition, although advocated for it. In the paper by Elishakoff (1995) the following possible definition was suggested:

$$R = 1 - 1/n \qquad (6.103)$$

for the required safety factor always exceeds unity.

The definition implies that

$$P_f n = 1 \qquad (6.104)$$

where $P_f = 1 - R$ is a probability of failure, or unreliability. In other words, one could define a non-probabilistic unreliability as the reciprocal of the non-probabilistic safety factor. Still, such a definition may still not appeal to the practicing engineers: In order to achieve the level of non-probabilistic reliability of, say 0.9, one needs a safety factor of 10! It appears that this observation, as well as the drawbacks of other plausible definitions is the main reason why the engineering community may not adopt the non-probabilistic reliability.

One may ask a natural question: Why to propose a possible definition of the non-probabilistic reliability, if one is skeptical of its very usefulness? In this a "linguistic quibble", as it may appear of the first glance?

It is felt that this is not so; we followed here ancient sages who would strengthen the other point of you, by a more valid argument, but then still would disagree. With these observations, one must strongly disagree with the radical opinion of Good (1996) maintaining that the Ben-Haim's concept of non-probabilistic reliability is an "oxymoron."

In another paper, Good (1995) maintains, as the telling title of his study suggests, "reliability always depends on probability of course." Whereas, in general, this statement appears to be valid, still, the dependence of the reliability upon probability does not exclude its dependence on other concepts. Elishakoff and Colombi (1993) combined probabilistic analysis with convex modeling to obtain the upper and lower bounds of the displacement variance of a structure. Elishakoff and Li (1999) combined probabilistic and convex modeling techniques to derive upper and lower bounds of the probability of failure. What Good (1995, 1996) is against then, is the exclusively non-probabilistic treatment of the reliability by Ben-Haim (1994, 1997).

For six different levels of treatment of uncertainties in risk analysis one should consult the insightful article by Paté-Cornell (1996).

The "excursion" to non-probabilistic reliability is still useful, for it at least elucidates various possible approaches to the problem of reliability. As Berg (1961) maintained, the "reliability is problem number one." If so, various possible hybrid approaches to it may be pursued, even if the professional consensus presently accepts only a single one.

More recently, Ben-Haim (2001) discusses the information-gap uncertainty:

"Info-gap models quantify uncertainty as a size of the gap between what is known and what could be known."

Here, immediately, numerous questions arise on why the information that "could be known" is not presently "known"? Is this due to negligence of the particular designer or of the entire profession which did not think it was necessary to "know" what "could be known", or the price involved with the obtaining the information that "could be known" is too heavy presently, or because this needed information is not released by appropriate firms and companies? It appears that such an analysis, if defined as above, should include an answer to the posed question. Ben-Haim (2001) states:

"Info-gap models of uncertainty are particularly useful when data on the uncertainties are quite limited."

One needs to know how "limited" the information should be to utilize info-gap models. The following pertinent questions arise: Why for the data that is available and considered as appropriate for the info-gap models, one cannot use the probabilistic modeling? Why not to utilize in such circumstances the fuzzy sets based approaches? Is there a difference between the convex modeling of uncertainty (Ben-Haim and Elishakoff 1990) and the information-gap uncertainty? Is the "information-gap uncertainty" a new terminology or a new methodology? As we see there are plenty of questions, demanding further research and answers.

Media and engineering profession simultaneously talk about the "age of information", "information highway" and "information overloading." Why is that in the engineering profession we are still lacking information on probabilistic distributions of loads or properties of the materials? The former Vice President of U.S.A. Al Gore (1995) wrote:

"…we will connect and provide access to the National Information Infrastructure for every classroom, every library, and every hospital and clinic in the entire United States of America."

Engineers are dreaming about the time when the international data banks of the material properties will be available for those who want to check validity of probabilistic, fuzzy-sets based or convex analyses. One should mention that some steps are already undertaken in this direction (see, e.g., Arbocz 1982; Arbocz and Abramovich 1979; Singer et al, 1978; Scherrer and Schuëller, 1988; Cooke, Bedford, Meilijson and Meester, 1993; Cooke, 1996).

6.6   Concluding Comments on How to Treat Uncertainty in a Given Situation

This chapter reviews some pertinent questions associated with uncertainty modeling in analysis of structures. It gives a critical appraisal of the probabilistic method and describes a new, non-probabilistic philosophy. The former is valid when plentiful information (probability densities) is available on the uncertain quantities involved. The latter, convex modeling is appropriate when the existing data is scarce and no valid probabilistic models can be constructed. Both of these approaches deal with different facets of uncertainty treatment. Theory of probability and random processes is not the only way to deal with uncertainty. Indeed, as Freudenthal (1956) notes

"ignorance of the cause of variation does not make such variation random."

These two methods appear to successfully complement each other, to make useful judgments based on reliably available experimental information. New safety factor based on convex modeling is introduced.

It is apropos to quote from Benjamin and Cornell (1970):

"How engineer chooses to treat uncertainty in a given situation depends upon the situation. If the degree of variability is small, *and* if the consequences of any variability will not be significant the engineer may choose to ignore it by simply assuming that the variable will be equal to the best available estimate. This estimate might be the average of the number of past observations. This is typically done, for example, with the elastic constants of materials and the physical dimensions of many objects.

If, on the other hand, uncertainty is significant, the engineer may choose to use a "conservative estimate" of the factor. This has often been done, for example, by setting "specified minimum" strength properties of materials…Many questions arise in the practice of using conservative estimates. For example:

How can engineer maintain consistency in their conservatism from one situation to another? For example, separate professional committees set the specifies minimum bending strength of wood."

It was demonstrated by Elishakoff, Li and Starnes (2001) that in some situations there may be *no difference* between the probabilistic and convex models of uncertainty. In certain situations both approaches tend to yield close numerical results. If this intriguing statement is correct the following conclusions can be made:

(a) one can continue doing the uncertainty analysis of one's choise.

(b) Pragmatic approach would suggest that the analysis that leads to less calculations, should be preferred. Naturally, the convex modelling does not involve the probability densities. But one can pose valid questions or finding the bounding ellipsoids containing the experimental information.

Still, one has to bear in mind the idea propagated by Thoft-Christensen and Baker (1982):

"Upper limits to the individual loads and lower limits to material strength are not easily identified in practice, e.g. building occupancy loads, wind loads, the yield stress of steel, the cube or cylinder stress of concrete."

It appears instructive to conclude the discussion on non-probaibilistic approaches on the safety factors by a quote from Bodner (2003):

"I think it is generally understood that the concept of Factor of Safety (FS) in a measure of the difference between the normal operating conditions (or the maximum in some sense) of a system and the minimum conditions for which the system cannot function. That could be due to many factors such as excessive deformations or cessation of load carrying ability due to inadequacies of materials or structure or to extreme loadings. The determination of the appropriate FS is part of the design process and is usually specified in codes and design handbooks. Of course, the recommended FS is different for various applications such as aircraft, buildings, bridges, underwater vehicles, etc. In many cases, the specific FS depends on the design parameters.

A particular meaning of FS suggested by Plasticity investigators is that the FS of a structure is the collapse condition (load) divided by a geometrically similar load that would initiate yield at some point in the structure. This concept has been generalized and has been used in practical applications.

I know there is much interest in formulating FS based on statistical consideration and various criteria for "probability of failure" have been proposed. These may be useful in some cases. From my observations, it seems that failures are usually due to a particular defect or anomaly in the system or loading condition that was not taken into account in the design. In modern times, misuse of a computer program has led to problems."

Works in this direction are those by Drucker, Greenberg and Prager (1951), Pell (1952), Mendelson (1968), Bushnell (1990), Kamenjarzh and Weichert (1992), Kamenjarzh (1996), Gross-Weege (1997), and others. In this connection, Bushnell (2003) notes:

"One may ask, "If one performs the collapse analysis, does not he in essence determine the exact value of the safety factor?" A quick answer is "No", because a safety factor can cover for a multiple of sins: unknown imperfections, unknown material properties, etc. In principle, if one knows exactly what the structure and its loading and boundary conditions are, one could do an "exact" finite element analysis and compute the collapse load. Then one could do some sort of approximate analysis, say a bifurcation buckling analysis of the idealized perfect structure with nominal material properties and boundary conditions and loading. A factor of safety could then be computed as the ratio:

$$\frac{\text{collapse load of actual structure}}{\text{bifurcation buckling of nominal system}}$$

Also, one might assign different factors of safety corresponding to differente phenomena, such as $fs(1)$ for stress, $fs(2)$ for buckling, $fs(3)$ for vibration, etc. The various factors of safety, fs, would depend upon how well known the phenomena is and how sensitive the predictions are to the variations of the input data."

For further details the reader may consult the paper by Bushnell (1990).

Chapter 7

Stochastic Safety Factor by Birger and Maymon

"Statistical safety factor retains its conditional importance as a criterion for comparison of newly developed item and the existing experience, characterizing the reliability. Its basic advantage over the usual (deterministic) safety factor consists in the fact, that the comparison is brought to more homogeneous conditions…"

I.A. Birger (1975)

"It can be assumed that in the near future probabilistic design requirements will be introduced to the aircraft industry, as has already been done in some countries for civil engineering structures. Thus there is a reason for an intermediate procedure which will bridge the gap between the deterministic safety factor based approach and the probabilistic methods."

G. Maymon (2002)

"….the appearance of modern computers and stress measuring methods now permit to determine stresses with an enormous precision… But up to now, no one has succeeded to get rid of the safety coefficient."

J. Kogan (1985)

In this chapter we will discuss, at some detail, the concept of stochastic safety factor, independently introduced by Birger (1970) and Maymon (2002). It differs for the central safety factor, extensively discussed in Chapters 3-5, and constitutes one of the four possible definitions of the safety factor in the probabilistic context, as briefly exposed in Section 3.2.

7.1   Introductory Comments

It appears instructive here to quote from Bolotin (1968):
"Engineering design, as a rule, bears a deterministic character. The condition of non-failure $\Sigma_y \geq \Sigma$, which may take place with some reliability R, is replaced by a deterministic condition

$$n = \frac{\sigma_{y,design}}{\sigma_{design}} \geq n_{code} \tag{7.1}$$

or

$$\sigma_{design} \leq \frac{\sigma_{y,design}}{n_{code}} \tag{7.2}$$

146

The design value of the load σ_{design} and the design of the value of the strength $\sigma_{y,design}$ are chosen in some degree, arbitrarily: These may be mathematical expectations of maximal (minimal) values. Once the design values σ_{design} and $\sigma_{y,design}$ are established, *normative safety factor* s_{code} is chosen in such manner, that from conditions (7.1) and (7.2) the condition $\sigma_y \geq \sigma$ would follow with reliability R. This shows the interconnectedness of parameters in Eqs. (7.1) and (7.2)" [The quote was adjusted by the notations adopted in this book].

7.2   Definition of Stochastic Safety Factor

Consider a structural component with yield stress Σ_y subjected to a stress Σ. Both Σ_y and Σ themselves can each constituite functions of many random variables, such as material proprieties, dimension and external loads. Consider the case in which both are normally distribuited indipendent random variables. Let us use according to Maymon (2002) following values

$$E(\Sigma) = 1, \quad \sigma_\Sigma = 0.05$$
$$E(\Sigma_y) = 1.2, \quad \sigma_{\Sigma_y} = 0.06 \tag{7.3}$$

so that the coefficients of variation of each of the variables are equal

$$V_\Sigma = \frac{\sigma_\Sigma}{E(\Sigma)} = \frac{\sigma_{\Sigma_y}}{E(\Sigma_y)} = 0.05 \tag{7.4}$$

Deterministic analyst would identify each of the variables with its nominal value

$$\sigma_{y,no\,min\,al} = 1.2$$
$$\sigma_{no\,min\,al} = 1 \tag{7.5}$$

Thus, the classical safety factor amounts to

$$n_{classical} = \frac{1.2}{1.0} = 1.2 \tag{7.6}$$

For the so called "worst case design" the "minimum" value of the yield stress is defined as

$$\sigma_{y,min} = E(\Sigma_y) - 3\sigma_{\Sigma_y} = 1.2 - 3 \times 0.06 = 1.02 \tag{7.7}$$

The "maximum" value of the stress is defined as

$$\sigma_{max} = E(\Sigma) + 3\sigma_\Sigma = 1 + 3 \times 0.05 = 1.15 \tag{7.8}$$

The safety factor corresponding to the deterministic "worst case" definition is obtained as follows

$$s_{worst} = \frac{\sigma_{y,min}}{\sigma_{max}} = \frac{1.02}{1.15} = 0.887 \tag{7.9}$$

a value that is unacceptable in an ordinary design.

Consider now the so-called stochastic safety factor

$$Q = \frac{\Sigma_y}{\Sigma} \tag{7.10}$$

Using it one can attempt to answer the following question, in Maymon's (2002a) terminology:

"What is the probability that a structure (with given uncertainties) has a factor of safety greater than a given value?"

The following section is devoted to answering this inquiry.

7.3 Implication of the Stochastic Safety Factor

We will call the random variable defined in Eq(7.10) the *Birger–Maymon safety factor*. Let us consider the implications of the Birger–Maymon safety factor in the context of this example. One can evaluate the cumulative distribution function (CDF) of the safety factor Q using analytical expression for the CFD of a ratio of two random variables (see, *e.g.* Elishakoff, 1983, 1999). In the study by Maymon (2002a) the CFD of Q was obtained using numerical integration of the analytical expressions in the above book, and also by using one of the commercially available computer programs (see, Millwater *et al*, 1992). The Figure 7.1 shows the CFD for the random variable Q. From this figure one concludes that there is a probability of 50% that the factor of safety of this design is lower than or equal to 1.2, corresponding to point A in Figure 7.1.

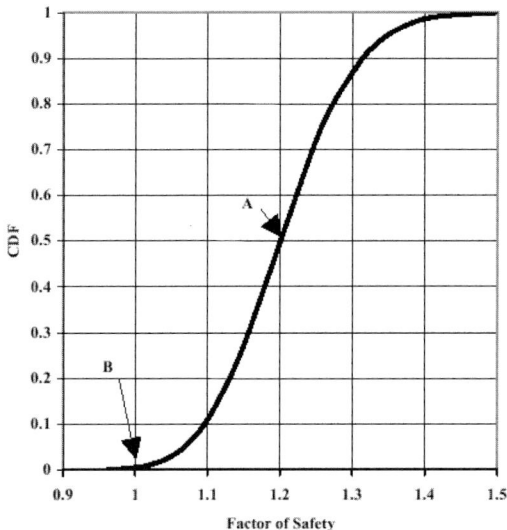

Fig. 7.1 Cumulative distribution function of the random safety factor

There is a probability of 0.00532 (point B of the figure) that the stochastic factor of safety will take an values less than unity. The probability of failure is $Prob(\Sigma_y < \Sigma)$, or simply, the $Prob(Q<1)$. Thus, the probability of failure is obtained from $F_Q(q)$ when $q=1$. Hence the probability of the failure of this structural component is 0.00532, and its reliability is 1-0.00532=0.99468.

Consider another case (Maymon, 2000a):

$$E(\Sigma_y) = 1.2, \quad \sigma_{\Sigma_y} = 0.12$$
$$E(\Sigma) = 1.0, \quad \sigma_{\Sigma} = 0.10 \tag{7.11}$$

Now the coefficient of variation equals 10% the nominal safety factor, as well as the central safety factor in this case equals 1.2. The classical worst case safety factor is obtained from calculating the following values:

$$\sigma_{y,min} = E(\Sigma_y) - 3\sigma_{\Sigma_y} = 1.2 - 3\times0.12 = 0.84$$
$$\sigma_{max} = E(\Sigma) + 3\sigma_{\Sigma} = 1.2 + 3\times0.1 = 1.5 \tag{7.12}$$

Thus,

$$n_{worst} = \frac{0.84}{1.5} = 0.56 \tag{7.13}$$

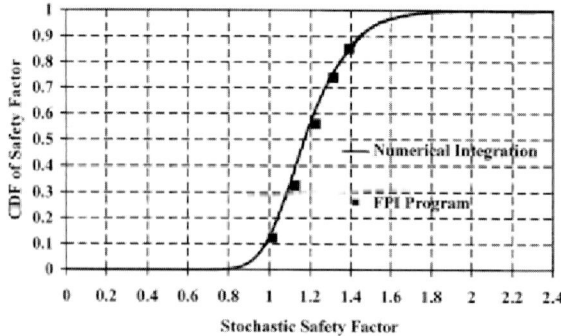

Fig. 7.2 Cumulative distribution function (CDF) of stochastic safety factor

Fig. 7.2 shows the CDF of Q obtained using numerical integration. We observe that the probability the safety factor is smaller than unity is 0.12. The probability that Q takes on values smaller than 1.2 is 0.56.

Maymon (2000a) performed parametric calculations for four values of $E(\Sigma_y)$, namely 1.2, 1.5, 2.0 and 3.0, while the mean square deviation σ_{Σ_y} was set to equal 0.12 in all above four case. The nominal and central safety factors for these case equal, respectively, 1.2, 1.5, 2.0 and 3.0. The probabilities to have the stochastic safety factor Q smaller than a given value are shown in Fig. 7.3.

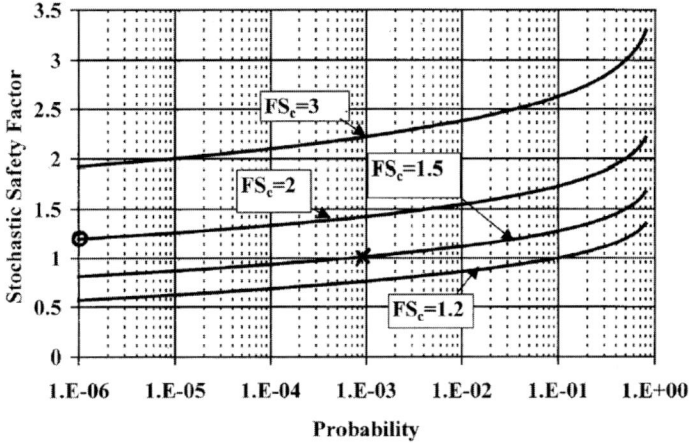

Fig. 7.3 Probability versus stochastic safety factor for four central safety factors (FS ≡ factor of safety)

We observe that for the second case (with central safety factor equal 1.5) the probability to have a stochastic safety factor smaller than unity is 9×10^{-4} (marked X in the figure). For the third case (central safety factor $n_1 = 2.0$) the probability to have a stochastic safety factor smaller than 1.2 is 1×10^{-6} (marked O).

Consider now the case in which both Σ and Σ_y have tree parameter Weibull distribution. This distribution is extensively used for many properties of materials. It is characterized by a minimum value γ_x for the variable X and by two constants δ and β parameters that can be determined by the values of the mean and variance of the distribution. Let the probabilistic moments and minimum values for Σ_y and Σ be given as follows

$$E(\Sigma_y) = 2, \quad \sigma_{\Sigma_y} = 0.5, \quad \gamma_{\Sigma_y} = 0.9$$
$$E(\Sigma) = 1.66667, \quad \sigma_{\Sigma} = 0.5, \quad \gamma_{\Sigma} = 0.8$$

(7.14)

The CDF of the stochastic Birger-Maymon safety factor Q calculated with the FPI/NESSUS program is shown in Fig. 7.4.

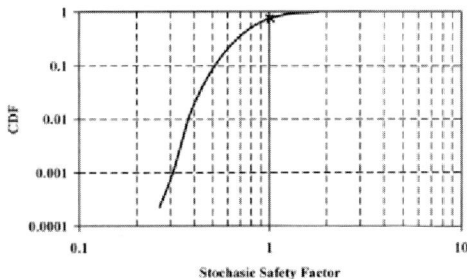

Fig. 7.4 Cumulative distribution function (CDF) of stochastic safety factor

It can be seen that in the calculated case the probability that the stochastic safety factor is smaller than unity is about 80% (designed by X in the figure) The reliability is calculated as a value

$$R = 1 - CDF\big|_{q=1} \tag{7.15}$$

implying that the cumulative probability distribution function at $q = 1$ equals the unreliability (probability of failure)of the structure. In this particular case $R \approx 0.2$, implying a poor design.

7.4   Cantilever Beam with Restricted Maximum Displacement

In this example, we treat a cantilever beam of length L, with a rectangular cross section. B denotes the width, while H indicates the depth. Young modulus is E; the beam is subjected to a tip force P. Design requires the tip displacement

$$Y_{tip} = \frac{4PL^3}{EBH^3} \tag{7.16}$$

not to exceed the maximum allowable value Y_{max}. All above quantities are random variables. The failure function is given by

$$g = Y_{max} - Y_{tip} = Y_{max} - \frac{4PL^3}{EBH^3} \tag{7.17}$$

where the failure region is given by the inequality $g \le 0$. The numerical data are as follows

$$
\begin{aligned}
L &= [149,151] \ mm \\
B &= [9.9,10.1] \ mm \\
H &= [4.9,5.1] \ mm \\
E &= [19000,21000] \ kgf / mm^2 \\
P &= [9.1,11.1222] \ kgf \\
Y_{max} &= [6.2,6.8] \ mm
\end{aligned}
\tag{7.18}
$$

These parameters are assumed to be normally distributed, and it is supposed that the tolerances represent ±3 standard deviations. Hence the mean values and the associated standard deviations are, respectively,

$$
\begin{aligned}
E(L) &= 150, & \sigma_L &= 0.3333 \\
E(B) &= 10, & \sigma_B &= 0.03333 \\
E(H) &= 5, & \sigma_H &= 0.03333 \\
E(E) &= 21{,}000, & \sigma_E &= 666.667 \\
E(P) &= 10.1111, & \sigma_P &= 0.3370 \\
Y_{max} &= 6.5, & \sigma_{Y_{max}} &= 0.1
\end{aligned}
\tag{7.19}
$$

A "nominal", deterministic analysis refers to the mean values of the parameters. The mean tip displacement is obtained from Eq. (7.14) 5.2 mm. Hence the nominal safety factor equals

$$s_{nominal} = \frac{6.5}{5.2} = 1.25 \qquad (7.20)$$

Let us perform the so-called "worst case" analysis. To conduct it we use the upper values for the quantities P and L appearing in the numerator of Eq. (7.14) while using the lower bounds for B, H and E, since these appear in the denominator. The maximum value for the tip displacement is, therefore

$$Y_{tip,max} = \frac{4P_{max}L_{max}^3}{E_{min}B_{min}H_{min}^3} = 6.92 \ mm \qquad (7.21)$$

The lowest allowable maximum tip displacement is (6.5-0.3=) 6.2 mm. The worst case safety factor equals

$$s_{worst\ case} = \frac{6.2}{6.92} = 0.896 \qquad (7.22)$$

and, thus, constitutes an unacceptable value in the conventional analysis.

Maymon (2002a) also conduced the regular probabilistic analysis, using FPI Probabilistic Structural analysis program (Millwater et al, 1992), via the first order analysis. Probability of failure equals 0.000013547 while its complement, the reliability is 0.999986453.

The Birger-Maymon safety factor equals

$$Q = \frac{Y_{max}}{4PL^3 / EBH^3} \qquad (7.23)$$

The calculations for the probability distribution function of this index are given in Fig. 7.5.

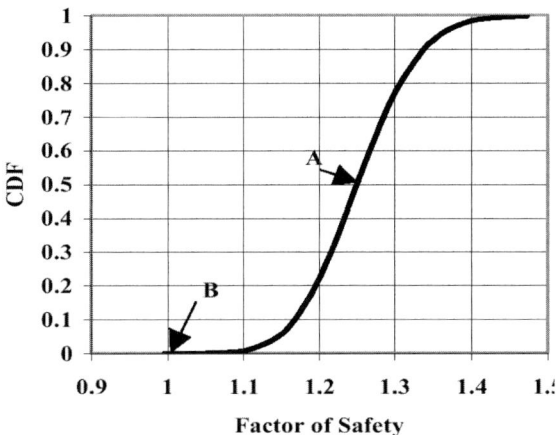

Fig.7.5 Probability that the safety factor does not exceed value 1.25 equals 0.5

It is readily observed that the probability that the beam has a Birger-Maymon safety factor not exceeding the nominal value, namely 1.25, is 50% and is denoted by point A in Fig. 7.5. The probability that the safety factor does not exceed unity equals 0.000013547, represented by the point B in the figure. This, as it should be, equals the probability of failure.

7.5   Concluding Comments

It appears instructive summarize this chapter by quoting Maymon (2002a):

> "The stochastic safety factor may answer the question "what is the probability that a given structure (with given uncertainties) has a safety factor equal to or smaller than a given value." The use of this safety factor can incorporate the structure into the system design, by applying a reliability value to the structure in the same way it is done for any other sub-system of the complete design. Commercially available probabilistic structural analysis computer programs can support the computations of stochastic safety factors…the use of probabilistic structural analysis is a necessary step in the evaluation of structural analysis towards more efficient design, hence the use of the stochastic safety factor is recommended."

Chapter 8

Safety Factor in Light of the Bienaymé-Markov and Chebychev Inequalities

"The Chebychev inequality gives a conservative but quantitative correspondence between safety limit and the lower bound of safety probability. Such inequalities will play an important role even when a certain one-to-one correspondence between safety limit and safety probability is found. Only safety limit will enter the actual design process."

H.L. Su (1961)

"…The distribution of the statistical variables involved in….analysis (loads, resistance, times of failure, defects, etc.) are neither known nor obtainable by direct statistical inference within the probability range that is significant for safety and reliability analysis."

A.M. Freudenthal (1975)

"Traditionally, the reliability of engineering systems is achieved through the use of factors or margins of safety…The traditional approach is difficult to quantify and lacks the logical basis for addressing uncertainties; consequently, the level of safety or reliability cannot be assessed quantitatively."

A.H-S. Ang and W.H. Tang (1984)

"The conventional [deterministic] way of assessing the design is not adequate with respect to the statistical scatter of the design parameters."

O. Vinogradov (1991)

"We are having fun developing novel insights as well as ingenious, efficient methods of analysis. Let us have more of it. Some insights will be fruitful, and some methods will become powerful every-day tools."

E. Rosenblueth (1991)

We have amply seen in previous chapters that in order to relate the reliability and the safety factor, one needs the knowledge of the probability densities of the involved random variables. Often, such information is missing. It is shown in this chapter that the celebrated inequalities by Irenée-Jules Bienaymé (1796-1878), Andrei Andreevich Markov (1856-1922) and Pafnutii Lvovich Chebychev (1821-1894) may come to help in finding bounds on the desired probabilities and safety factors. Short biographical notes about Bienayme and Chebychev are given in the Appendix B.

8.1 Bienaymé-Markov Inequality

Let X denote a random variable that takes on only positive values, that is

$$f_X(x) = 0 \text{ for } x < 0 \tag{8.1}$$

The Bienaymé-Markov inequality maintains that for a positive constant α, the following statement is valid:

$$Prob(X \geq \alpha) \leq \frac{E(X)}{\alpha} \tag{8.2}$$

Let us first stress that if this inequality is correct, it states that the probability of excedance, $Prob(X \geq a)$ can be found by the quantity $E(X)/\alpha$, i.e. depends only upon the mean value $E(X)$ of the random variable X. Proof of this inequality is extremely simple; indeed, according to the definition of the mathematical expectation

$$E(X) = \int_{-\infty}^{\infty} x f_X(x) dx \tag{8.3}$$

The integration domain $(-\infty, \infty)$ can be divided as two domains $(-\infty, \alpha)$ and (α, ∞). Thus

$$E(X) = \int_{-\infty}^{\alpha} x f_X(x) dx + \int_{\alpha}^{\infty} x f_X(x) dx \tag{8.4}$$

Since all values of x are taken as positive then both the integrals on right-hand-side of the above equation will always be positive. If the first term in the right-hand-side is dropped, we get the following obvious inequality

$$E(X) \geq \int_{\alpha}^{\infty} x f_X(x) dx \tag{8.5}$$

Furthermore, as the integration variable x varies in the interval (α, ∞), the inequality will be strengthened if the variable x in the expression of the integrand in Eq. (8.4) will be replaced by the minimal value taken by x from the interval, namely α. Hence,

$$E(X) \geq \int_{\alpha}^{\infty} \alpha f_X(x) dx = \alpha \int_{\alpha}^{\infty} f_X(x) dx \tag{8.6}$$

We observe that in order to evaluate $Prob\ (X \geq \alpha)$ we should integrate the probability density over the values that exceed α. In other words,

$$E(X) \geq \alpha\ Prob(X \geq \alpha) \tag{8.7}$$

which coincides with the Bienaymé-Markov inequality in Eq. (6.2).

8.2   Use of the Bienaymé-Markov Inequality for Reliability Estimation

Consider a uniform bar of constant cross-sectional area $a = 4$ mm^2 that is under the tensile force P with mean load $E(P) = 400$ *Newtons*. The yield stress is a deterministic quantity $\sigma_y = 250$ N/mm^2. We would like to use the Bienaymé-Markov inequality to estimate the probability of failure. The tensile stress, $\Sigma = P/a$ has the mean value

$$E(\Sigma) = \frac{E(P)}{a} = \frac{400\ N}{4\ mm^2} = 100\ N/mm^2 \tag{8.8}$$

The probability of failure is the probability that the actual stress exceeds the yield stress, σ_y. In accordance with the Bienaymé-Markov inequality, the probability of failure is estimated as

$$P_f = Prob(\Sigma > \sigma_y) \le \frac{E(\Sigma)}{\sigma_y} = \frac{100\ N/mm^2}{250\ N/mm^2} = 0.4 \tag{8.9}$$

This inequality may often yield rather rough estimates because *only* the mean value of the random variable is assumed as known. Indeed, knowledge that the probability of failure is less than or equal to 0.4 or that the reliability is not less than 0.6 may be insufficient. Based on such an estimate, one may not be in a position to give a "green light" to the actual use of the structure, especially if a high performance reliability, say 0.9 or greater is required.

Yet, if the yield stress constitutes 1000 N/mm^2 then the estimate of the probability of failure is

$$P_f = Prob(\Sigma > \sigma_y) \le \frac{E(\Sigma)}{\sigma_y} = \frac{100\ N/mm^2}{1000\ N/mm^2} = 0.1 \tag{8.10}$$

The knowledge that the upper bound of the probability of failure is 0.1 may allow utilization if the structure, if the permissible probability of failure is not less than 0.1.

For comparison purposes, let us consider the case when the probability density of the load is inferred from the experiments, and, thus, is known. Let P be a uniformly distributed random variable in the interval (200, 600) *Newtons*. Then the actual stress Σ is also a uniformly distributed random variable in the interval

$$\left(\frac{200}{4}, \frac{600}{4}\right) = (50,150)\ N/mm^2 \tag{8.11}$$

The mean value of this stress distribution equals

$$E(\Sigma) = \frac{50 + 150}{2} = 100\ N/mm^2 \tag{8.12}$$

If the yield stress is a deterministic quantity $\sigma_y = 250$ N/mm^2, the probability of failure vanishes

$$P_f = Prob(\Sigma > \sigma_y) = 0 \tag{8.13}$$

Indeed, the maximum value of the actual stress, 150 N/mm^2 is less than the yield stress. Thus, the structure never fails, and the failure probability is zero, whereas the bound provided by the Bienaymé-Markov inequality was 0.4, as Eq. (8.9) states. So, the estimate provided by the Bienaymé-Markov inequality was quite poor. If $\sigma_y = 140$ N/mm^2 Bienaymé-Markov inequality provides the following bound

$$P_f = Prob(\Sigma > 140) \le \frac{E(\Sigma)}{140} = \frac{100\ N/mm^2}{140\ N/mm^2} = 0.714 \tag{8.14}$$

The exact value for P_f equals

$$P_f = \int_{140}^{\infty} f_\Sigma(\sigma)d\sigma = \int_{140}^{150} f_\Sigma(\sigma)d\sigma \tag{8.15}$$

The probability density of the stress equals, if it is uniformly distributed in the interval (50, 150) *Newtons*,

$$f_\Sigma(\sigma) = \begin{cases} 0, & for\ \sigma < 50\ N/mm^2 \\ \dfrac{1}{150-50} = 0.01, & for\ \sigma \in (50,150)\ N/mm^2 \\ 0, & for\ \sigma > 150\ N/mm^2 \end{cases} \tag{8.16}$$

Substitution of Eq. (8.16) into Eq. (8.15) results in

$$P_f = \int_{140}^{150} 0.01 d\sigma = 0.1 \tag{8.17}$$

In this case too, the bound, 0.714, provided by the Bienaymé-Markov inequality in Eq. (8.14) is much greater than the actual value of 0.1 in Eq. (8.17).

If Σ is an exponentially distributed random variable with the probability distribution

$$F_\Sigma(\sigma) = \begin{cases} 0, & for\ \sigma < 0\ N/mm^2 \\ 1 - exp\left[\dfrac{-\sigma}{E(\Sigma)}\right], & for\ \sigma > 0\ N/mm^2 \end{cases} \tag{8.18}$$

and mean value $E(\Sigma) = 100\ N/mm^2$, then the probability of failure equals

$$P_f = Prob(\Sigma > \sigma_y) = 1 - Prob(\Sigma \le \sigma_y)$$

$$= 1 - F_\Sigma(\sigma_y) = 1 - \left\{1 - exp\left[-\frac{\sigma_y}{E(\Sigma)}\right]\right\} \tag{8.19}$$

$$= exp\left[-\frac{\sigma_y}{E(\Sigma)}\right]$$

or

$$P_f = exp\left[-\frac{250}{100}\right] = exp(-2.5) = 0.082 \tag{8.20}$$

which is much smaller than the bound provided by the Bienaymé-Markov inequality.

This underlines the importance of the knowledge of the probability densities of the involved variables. The bounds provided by the Bienaymé-Markov inequality should be used only when no information is available about the probability densities. In general, if one sets $\alpha = \sigma_y$ in Eq. (8.2) while X is set to equal the actual stress, Σ, the result is

$$Prob\left(\Sigma \geq \sigma_y\right) \leq \frac{E(\Sigma)}{\sigma_y} \qquad (8.21)$$

The quantity $E(\Sigma)/\sigma_y$ is naturally reciprocal to the central safety factor. Then the inequality in Eq. (8.18) is rewritten as follows

$$P_f \leq \frac{1}{n} \qquad (8.22)$$

We arrive at an interesting conclusion stating that for the element under random actual stress with material possessing the deterministic yield stress, the probability of failure is bounded by the reciprocal of the central safety factor: the greater the central safety factor; the less the bound on the probability of failure. This conclusion illustrates the intimate connection of the notion of probability of failure and the central safety factor. Elishakoff (2000) provided an illustration of the use of the Bienaymé-Markov inequality to estimate the probability of failure.

Consider now a reverse case in which the actual stress is deterministic but the yield stress is a random variable. We substitute in Eq. (8.2)

$$X = \Sigma_y , \alpha = \sigma \qquad (8.23)$$

and get

$$Prob\left(\Sigma_y \geq \sigma\right) \leq \frac{E(\Sigma_y)}{\sigma} \qquad (8.24)$$

We recognize the left-hand-side as the reliability, whereas the right-hand-side is a central safety factor. Thus

$$R \leq n \qquad (8.25)$$

If the central safety factor exceeds unity, for example $n = 1.5$, the information contained in Eq. (8.25) is trivial. Indeed, not only $R \leq 1.5$, but it cannot exceed unity! Eq. (8.25) makes sense only if $n < 1$. However, in such a case, the safety requirement is violated with respect to the mean quantities. Usually, one can assume that the designer must meet the safety requirement with respect to nominal, mean quantities, in addition to meeting the reliability requirement. We conclude that in this case the Bienaymé-Markov inequality does not provide an useful information.

8.3   Derivation of the Chebychev's Inequality

Chebychev's inequality states: If X is a random variable with the finite mean value $E(X)$ and variance σ_x^2, then for say any positive value m, the following inequality holds:

158

Pafnutii Lvovich Chebychev (born on May 16, 1821 in Okatovo, Russia, died on December 8, 1894 in St. Petersburg, Russia)

$$Prob\left[|X - E(X)| \geq m\right] \leq \frac{\sigma_X^2}{m^2} \qquad (8.26)$$

Let us first describe the implication of this inequality. It states that regardless of the shape of the actual probability density of the random variable X, the probability

$$Prob\left[E(X) - \varepsilon < X < E(X) + \varepsilon\right] \geq 1 - \frac{\sigma_X^2}{\varepsilon^2} \qquad (8.27)$$

That X takes values in the interval $\left[E(X) - \varepsilon, E(X) + \varepsilon\right]$ centered at the mean value $E(X)$ is close to unity, provided that

$$\frac{\sigma_X^2}{\varepsilon^2} \ll 1 \qquad (8.28)$$

In contrast to the Bienaymé-Markov inequality, the Chebychev's inequality involves both the mean value $E(X)$ as well as its variance σ_X^2. To prove this inequality consider the random variable

$$Y = \left[X - E(X)\right]^2 \qquad (8.29)$$

That naturally is a non-negative random variable. Thus for $\alpha = m^2$ in Eq. (8.2) we get

$$Prob\left(Y \geq m^2\right) \leq \frac{E(Y)}{m^2} \qquad (8.30)$$

Hence, in view of Eq. (8.26),

$$\text{Prob}\left\{[X - E(X)]^2 \geq m^2\right\} \leq \frac{E[X - E(X)]^2}{m^2} \tag{8.31}$$

We reserve that $[X\text{-}E(X)]^2 \geq m^2$ if and only if $|X\text{-}E(X)| \geq m$. Eq. (8.28) then reduces to

$$\text{Prob}\left\{|X - E(X)| \geq m\right\} \leq \frac{E[X - E(X)]^2}{m^2} \tag{8.32}$$

or

$$\text{Prob}\left\{|X - E(X)| \geq m\right\} \leq \frac{\sigma_X^2}{m^2} \tag{8.33}$$

which completes the proof. Its convenient to express n in terms of standard deviation σ_X

$$m = k\,\sigma_X \tag{8.34}$$

Then Eq. (8.33) becomes

$$Prob\left\{|X - E(X)| \geq k\sigma_X\right\} \leq \frac{1}{k^2} \tag{8.35}$$

Chebychev's inequality enables us to derive bounds when both the mean $E(X)$ and variance σ_X^2 of the probability distribution are known. Ross (1998) notes:

> "Of course, if the actual distribution were known, then the desired probabilities could be exactly computed and we would not need to resort to bounds."

8.4   Application of the Chebychev's Inequality: Mischke's Bound

Let us treat the safety factor as a random variable,

$$Q = \frac{\Sigma_y}{\Sigma} \tag{8.36}$$

We apply the Chebychev's inequality to the random safety factor by formally substituting $X \equiv Q$ in Eq. (8.35)

$$Prob\left\{|Q - E(Q)| \geq k\sigma_Q\right\} \leq \frac{1}{k^2} \tag{8.37}$$

In other words, the probability that the safety factor Q takes on values outside the band of $\pm k$ standard deviation's on either side of the mean is equal to or less than $1/k^2$.

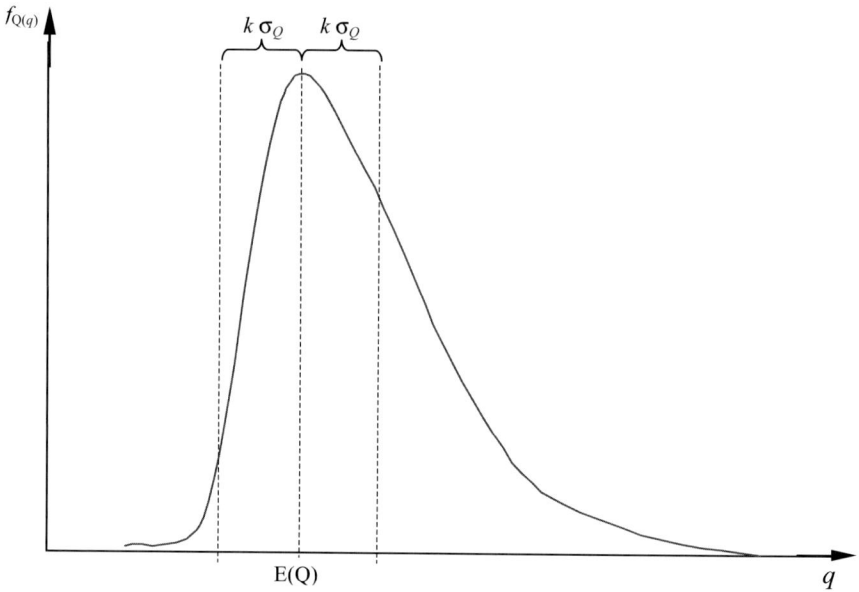

Fig. 8.1 The probability density function of random safety factor

In Fig. 8.1 the distance between the unit abscissa $q = 1$ and the distribution mean $E(Q)$ is $E(Q) - 1$ in units of q or $[E(Q) - 1]/\sigma_Q$ in units of standard deviation of Q. Mischke (1970) observed that the fraction of observations of Q that lies outside the range

$$E(Q) - k\sigma_Q \leq Q \leq E(Q) + k\sigma_Q \tag{8.38}$$

is not greater than $1/k^2$, where

$$k = \frac{E(Q) - 1}{\sigma_Q}. \tag{8.39}$$

Substituting the value into the Chebychev's inequality, we get

$$Prob\{|Q - E(Q)| \geq E(Q) - 1\} \leq \frac{1}{k^2} \tag{8.40}$$

Alternatively

$$Prob\{Q - E(Q) \geq E(Q) - 1 \bigcup Q - E(Q) \leq -[E(Q) - 1]\} \leq \frac{1}{k^2} \tag{8.41}$$

or

$$Prob\{Q \geq 2E(Q) - 1 \bigcup Q \leq 1\} \leq \frac{1}{k^2} \tag{8.42}$$

The events $Q \geq 2E(Q) - 1$ and $Q \geq 1$ are mutually exclusive. Therefore,

$$Prob\{Q \ge 2E(Q) - 1 \cup Q \ge 1\} = Prob\{Q \ge 2E(Q)\} + Prob(Q \ge 1) \qquad (8.43)$$

Thus Eq. (8.40) becomes,

$$Prob\{Q \ge 2E(Q)\} + Prob(Q \ge 1) \le \frac{1}{k^2} \qquad (8.44)$$

The inequality will not become weaker, if the first part is dropped, resulting in the following inequality

$$Prob(Q \ge 1) = P_f \le \frac{1}{k^2} \qquad (8.45)$$

Substituting the definition of k (8.39), and rewriting the equation we get,

$$P_f \le \frac{1}{k^2} = \frac{\sigma_Q^2}{[E(Q)-1]^2} = \frac{E(Q)^2}{[E(Q)-1]^2}\left(\frac{\sigma_Q}{E(Q)}\right)^2 \qquad (8.46)$$

Now substituting $P_f = 1 - R$ and solving for $E(Q)$ we get

$$E(Q) \ge \frac{1}{1 - \gamma_Q / \sqrt{1-R}} \qquad (8.47)$$

where γ_Q is the co-efficient of variation of the safety factor

$$\gamma_Q = \frac{\sigma_Q}{E(Q)} \qquad (8.48)$$

As Mischke (1970) notes, "Eq. (8.47) establishes the smallest value of $E(Q)$ which ensures that the fraction of R instances of observations of Q will fall with in the range $1 < E(Q) < E(Q) + [E(Q)-1]$. Less than $1 - R$ will fall outside the range. Included in all instances that fall without the range are all instances of failure, $Q < 1$ and some instances of non failure, $Q < E(Q) + [E(Q)-1]$."

Mischke (1970) also refers to the Camp-Meidell inequality: "Camp-Meidell theorem of statistics states that in a distribution exhibiting a single central tendency (unimodal with single extreme) and having a high order of contact with the abscissa at $\pm\infty$, the fraction of observations that will fall outside a band of $\pm k$ standard deviation on either side of the mean is equal to or less than $1/(2.25k^2)$. In other words, the probability of observing an instance of Q outside of $\pm k$ standard deviations from $E(Q)$ is $1/(2.25k^2)$ or less. In this case from the figure the distance between unit abscissa and the distribution mean is $E(Q)$ -1 in units of Q or $(E(Q) -1)/\sigma_Q$ in units of standard deviation of Q:

$$E(Q) - k\sigma_Q \le Q \le E(Q) + k\sigma_Q \qquad (8.49)$$

is $1/k^2$, where

$$k = \frac{E(Q)-1}{\sigma_Q} \qquad (8.50)$$

Substituting the value into the Chebychev's inequality, we get

$$Prob\{|Q - E(Q)| \geq E(Q) - 1\} \leq \frac{1}{2.25k^2} \tag{8.51}$$

Alternatively

$$Prob\{Q - E(Q) \geq E(Q) - 1 \cup Q - E(Q) \leq -[E(Q) - 1]\} \leq \frac{1}{2.25k^2} \tag{8.52}$$

or

$$Prob\{Q \geq 2E(Q) - 1 \cup Q \geq 1\} \leq \frac{1}{2.25k^2} \tag{8.53}$$

The events $Q \geq 2E(Q) - 1$ and $Q \geq 1$ are mutually exclusive. Therefore,

$$Prob\{Q \geq 2E(Q) - 1 \cup Q \geq 1\} = Prob\{Q \geq 2E(Q)\} + Prob(Q \geq 1) \tag{8.54}$$

Thus Eq. (8.40) becomes,

$$Prob\{Q \geq 2E(Q)\} + Prob(Q \geq 1) \leq \frac{1}{2.25k^2} \tag{8.55}$$

The inequality will not become weaker, if the first part is dropped, resulting in the following inequality

$$Prob(Q \geq 1) = P_f \leq \frac{1}{2.25k^2} \tag{8.56}$$

Substituting the definition of k (8.39), and rewriting the equation we get ,

$$P_f \leq \frac{1}{2.25k^2} = \frac{\sigma_Q^2}{2.25[E(Q) - 1]^2} = \frac{E(Q)^2}{2.25[E(Q) - 1]^2} \left(\frac{\sigma_Q}{E(Q)}\right)^2 \tag{8.57}$$

Now substituting $P_f = 1 - R$ and solving for $E(Q)$ we get

$$E(Q) \geq \frac{1}{1 - \gamma_U / 1.5\sqrt{1 - R}} \tag{8.58}$$

where γ_Q is the coefficient of variation of the safety factor

$$\gamma_Q = \frac{\sigma_Q}{E(Q)} \tag{8.59}$$

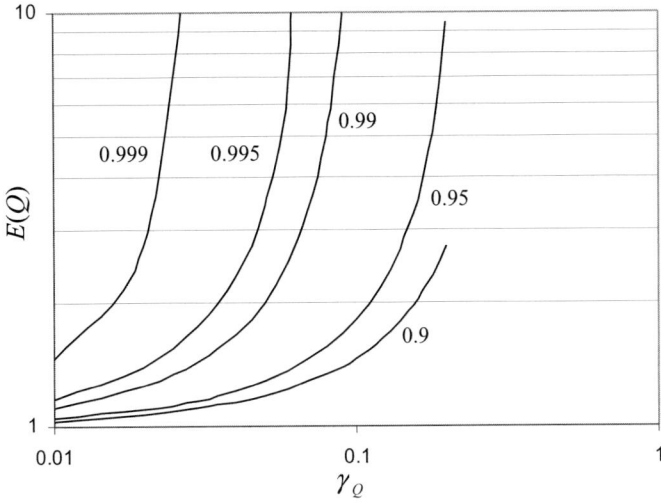

Fig. 8.2 The relationship between the mean safety factor and the coefficient of variation γ_Q obtained via the Chebychev inequality

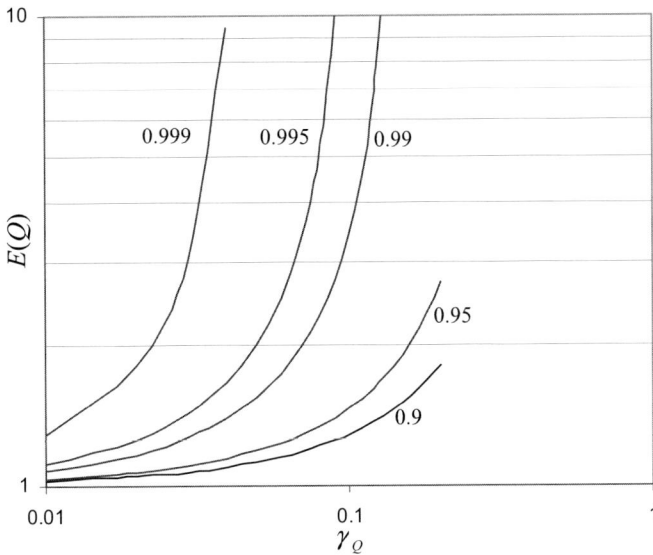

Fig. 8.3 The relationship between the mean safety factor and the coefficient of variation γ_Q obtained via the Camp-Meidell inequality

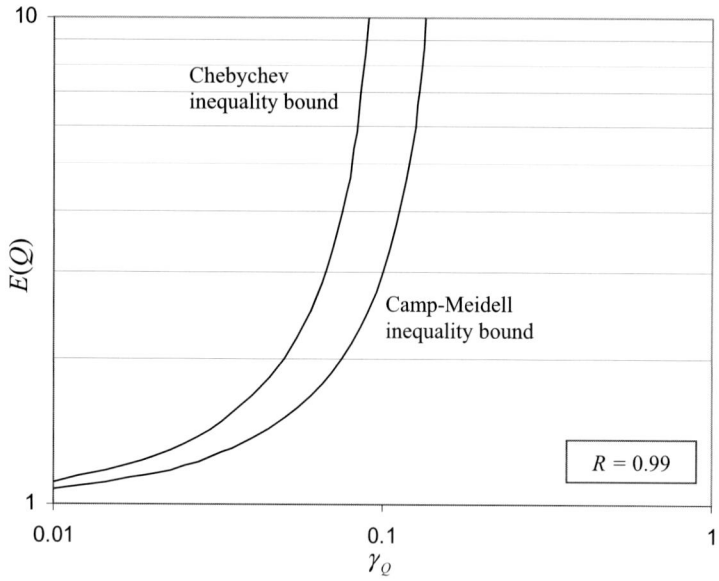

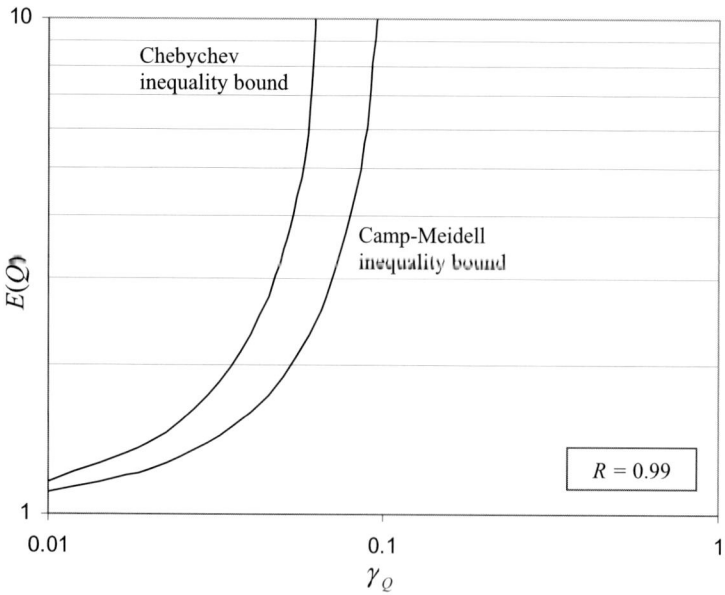

Fig. 8.4 Comparision graphs of Chebychev inequality bound and Camp-Miedell inequality bound for various values of reliability

8.5 Application of the Chebychev's Inequality by My Dao-Thien and Massoud

This section follows closely the paper by My Dao-Thien and Massoud (1974). Fig. 6.1 represents the probability density function $f_Q(q)$, in which (a) represents a value of the random quantity Q greater than the mean value $E(Q)$. According to the Chebychev inequality, the probability of obtaining a value Q within the range $|Q-a| \leq \varepsilon$ is estimated as

$$P(|Q-a| \leq \varepsilon) \geq 1 - \frac{E[(Q-a)^2]}{\varepsilon^2} \qquad (8.60)$$

For the values

$$a = kE(Q) \ , \ a - \varepsilon = 1 \qquad (8.61)$$

we get

$$E[(Q-a)^2] = E[(Q - kE(Q))^2]$$

$$= E(Q^2) - 2k[E(Q)]^2 + k^2[E(Q)]^2 \qquad (8.62)$$

$$= [E(Q)]^2[v_Q^2 + (1-k)^2]$$

where, v_Q is the coefficient of variation and k is an arbitrary constant greater than one.

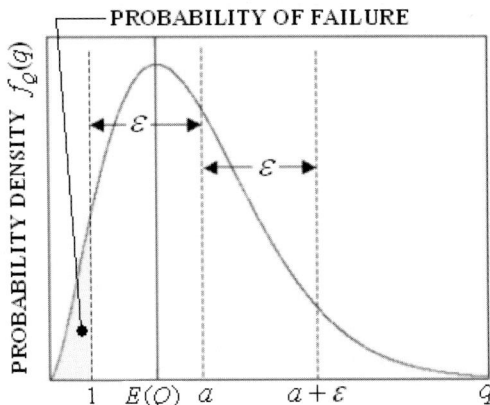

Fig. 8.5 The probability density of the factor of safety

Eq. (8.60) can be written in the following form,

$$P(1 \leq Q \leq 2kE(Q)-1) \geq 1 - \frac{[E(Q)]^{-2}\left[v_Q^{2}+(1-k)^{2}\right]}{[kE(Q)-1]^{2}} \qquad (8.63)$$

The probability of no failure of a component defines its reliability, R, and is given by $R = P(q > 1)$. From $R = P(q > 1)$ and Eq. (8.63) we write,

$$R \geq 1 - \frac{[E(Q)]^{-2}\left[v_Q^{2}+(1-k)^{2}\right]}{[kE(Q)-1]^{2}} \qquad (8.64)$$

for any value of k greater than one. The equality sign represents the lower bound of the reliability. The highest lower bound is obtained when the value Z

$$\frac{[E(Q)]^{-2}\left[v_Q^{2}+(1-k)^{2}\right]}{[kE(Q)-1]^{2}} = Z \qquad (8.65)$$

attains minimum. Differentiating Z once with respect to k and equating the result to zero, we get the critical value of the constant k:

$$k^{*} = \frac{E(Q)(v_Q^{2}+1)-1}{[E(Q)-1]} \qquad (8.66)$$

Differentiating once more with respect to k and substituting the value of k^{*} given in Eq. (8.66), we get,

$$\left.\frac{d^{2}Z}{dk^{2}}\right|_{k=k^{*}} = \frac{2[E(Q)]^{2}[E(Q)-1]^{4}}{\left\{[E(Q)]^{2}v_Q^{2}+[E(Q)-1]^{2}\right\}^{3}} \qquad (8.67)$$

The right hand side of Eq. (8.67) is always positive, indicating that k^{*} minimizes Z. Substituting k^{*} for k in Eq. (8.66) we get

$$R \geq \frac{[E(Q)]^{-2}v_Q^{2}}{\left\{[E(Q)]^{2}v_Q^{2}+[E(Q)-1]^{2}\right\}} \qquad (8.68)$$

where the equality sign represents the highest lower bound of reliability. Eq. (8.68) can be rewritten in the form

$$E(Q) \geq \frac{1}{1-v_Q\sqrt{\dfrac{R}{1-R}}} \qquad (8.69)$$

Eq. (8.69) establishes the smallest value of the mean factor of safety, which insures that the probability of finding Q in the range $1 < Q < [2k^* E(Q) - 1]$ will be R.. The lower bound of the factor of safety is given by utilizing the equality sign in Eq. (8.69). The relationship between the mean factor of safety $E(Q)$, the coefficient of variation v_Q and the reliability R as derived from Eq. (8.69) are shown in Fig. 8.6. In Eq. (8.69), an approximate expression of the reliability as a function of the parameters $E(Q)$ and v_Q determining the probability of density function $f_Q(q)$ is given. For most engineering applications the statistical values defining $f_{\Sigma_y}(\sigma_y)$ and $f_\Sigma(\sigma)$ must be obtained from sufficient observations. A practical expression of the reliability as a function of these statistics is required. The central factor of safety s_1, which is a reinterpretation of the classical deterministic factor of safety, could also be a parameter in the reliability expression instead of the mean factor of safety $E(Q)$, which appears in Eq. (8.69).

From $R = P(q > 1)$, the mean value $E(Q)$ is given for independent random quantities Σ_y and Σ,

$$E(Q) = E\left(\frac{\Sigma_y}{\Sigma}\right) = E(\Sigma_y) \cdot E\left(\frac{1}{\Sigma}\right) \qquad (8.70)$$

The mean value $E(1/\Sigma)$ is derived from the approximate mathematical relation (Elishakoff, 1983, 1999):

$$E[g(X)] \cong g[E(X)] + g''[E(X)]\frac{\sigma_X^2}{2} \qquad (8.71)$$

where, $g(X)$ is a function of a random quantity X with a mean value $E(X)$ and a standard deviation σ_X^2.

For $g(X) = g(E) = 1/\Sigma$, Eq. (6.42) gives,

$$E\left(\frac{1}{\Sigma}\right) \cong \frac{1}{E(\Sigma)} + \frac{\sigma_\Sigma^2}{[E(\Sigma)]^3} \qquad (8.72)$$

From Eqs. (8.70) and (8.72) we obtain,

$$E(Q) \cong E(\Sigma_y)\left\{\frac{1}{E(\Sigma)} + \frac{\sigma_\Sigma^2}{[E(\Sigma)]^3}\right\} \cong s_1(1 + v_\Sigma^2) \qquad (8.73)$$

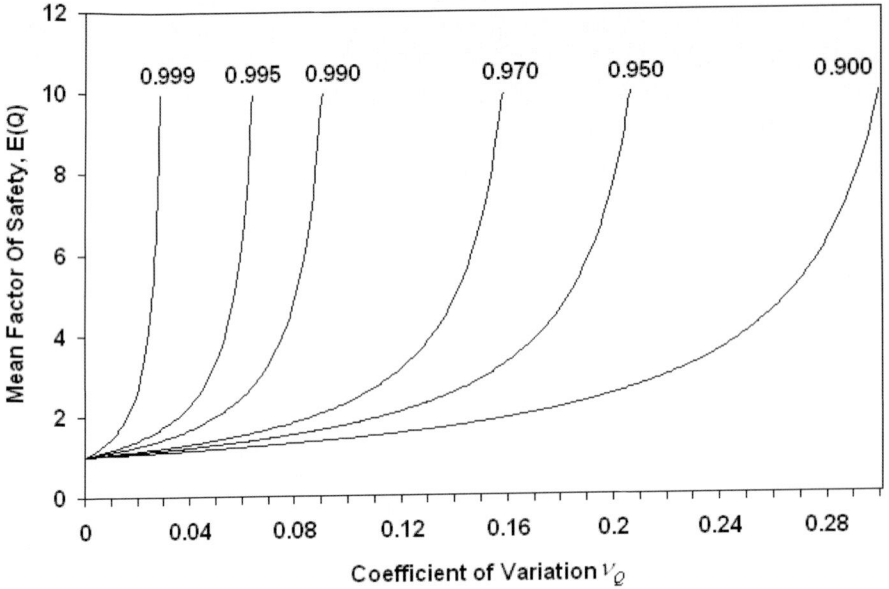

Fig. 8.6 Relationship between the mean factor of safety $E(Q)$ and coefficient of variation V_Q, for various levels of reliability R

We can also write, for statistically independent Σ and Σ_y,

$$E(Q^2) = E(\Sigma_y^{\,2}) \cdot E\left(\frac{1}{\Sigma^2}\right) \tag{8.74}$$

from Eqs. (8.71) and (8.74) for $g(X) = g(\Sigma) = 1/\Sigma^2$ we get,

$$E(Q^2) \cong E(\Sigma_y^{\,2})\left\{\frac{1}{\Sigma^2} + 3\frac{\sigma_\Sigma^{\,2}}{[E(\Sigma)]^4}\right\} \tag{8.75}$$

therefore,

$$\sigma_Q^{\,2} + [E(Q)]^2 \cong \left\{\sigma_{\Sigma_y}^{\,2} + [E(\Sigma_y)]^2\right\}\left\{\frac{1}{\Sigma^2} + 3\frac{\sigma_\Sigma^{\,2}}{[E(\Sigma)]^4}\right\} \tag{8.76}$$

Dividing Eq. (8.76) by $[E(Q)]^2$ and substituting from Eq. (8.73) we get,

$$V_Q \cong \frac{\left[\left(1 + V_{\Sigma_y}^{\,2}\right)\left(1 + 3V_\Sigma^{\,2}\right) - \left(1 + V_\Sigma^{\,2}\right)^2\right]^{1/2}}{\left(1 + V_\Sigma^{\,2}\right)} \tag{8.77}$$

Substituting in Eqs. (8.64) and (8.66) from equations (8.73) and (8.77) we get,

$$R \geq 1 - \frac{n_1^2 \left[\left(1+v_{\Sigma_y}^2\right)\left(1+3v_\Sigma^2\right) - k^*\left(k^* - 2\right)\left(1+v_\Sigma^2\right)^2\right]^{1/2}}{\left[k^* q_1\left(1+v_\Sigma^2\right) - 1\right]^2} \tag{8.78}$$

where

$$k^* = \frac{n_1\left[\left(1+v_{\Sigma_y}^2\right)\left(1+3v_\Sigma^2\right)/\left(1+v_\Sigma^2\right)\right] - 1}{\left[s_1\left(1+v_\Sigma^2\right) - 1\right]} \tag{8.79}$$

The equality sign in Eq. (8.79) yields the lower bound of the reliability as a function of the central factor of safety n_1 and the parameters defining the probability density functions $f_{\Sigma_y}(\sigma_y)$ and $f_\Sigma(\sigma)$.

From Eqs. (8.69), (8.73) and (8.77) we express n_1

$$n_1 \geq \frac{1}{1 - \left\{\left[\frac{\left[\left(1+v_{\Sigma_y}^2\right)\left(1+3v_\Sigma^2\right) - \left(1+v_\Sigma^2\right)^2\right] r}{1-r}\right]^{0.5} - v_\Sigma^2\right\}} \tag{8.80}$$

where r is the required reliability level. Neglecting higher orders of $v_{\Sigma_y}^2$ and v_Σ^2 the foregoing equation can be simplified as

$$n_1 \geq \frac{1}{1 - \left\{\left[\frac{\left(v_{\Sigma_y}^2 + v_\Sigma^2\right) \cdot r}{1-r}\right]^{0.5} - v_\Sigma^2\right\}} \tag{8.81}$$

Eq. (8.81) gives the smallest value of the central factor of safety with at least a probability R of finding the actual factor of safety Q in the range $\left|1 < Q < \left\{2k^* s_1\left(1+v_\Sigma^2\right) - 1\right\}\right|$.

DESIGN NOMOGRAM

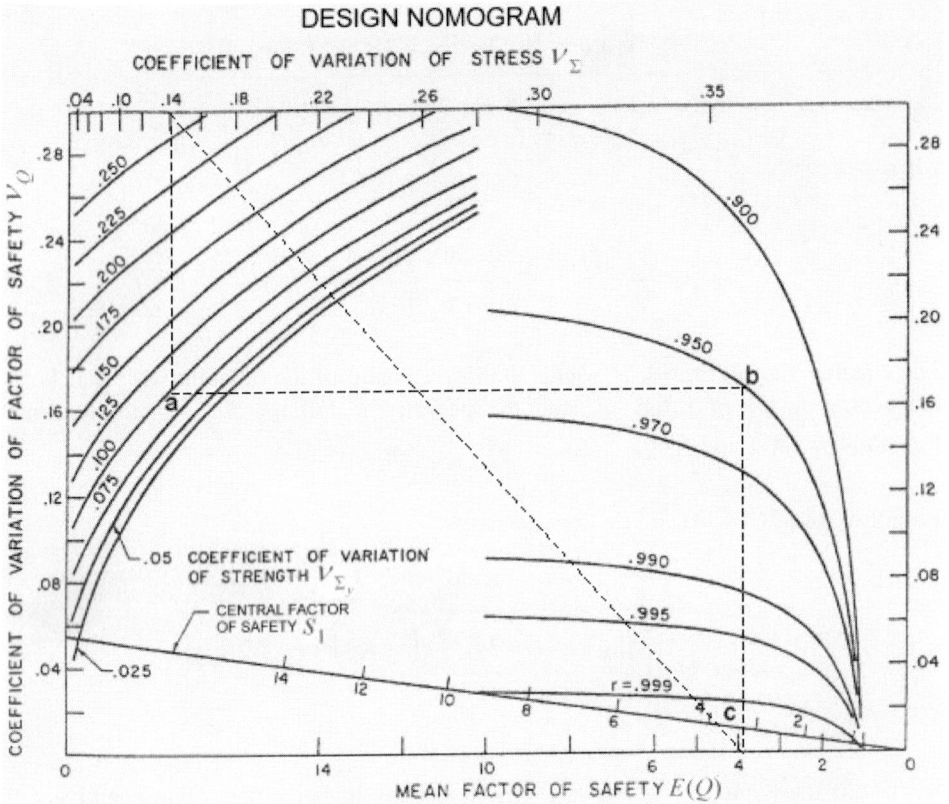

Fig. 8.7 Design nomogram

For practical applications, Eq. (8.81) is used to draw a design monogram, given in Fig. 8.7. Note that in Fig. 8.7 three are two horizontal axes: The lower one associates with the mean factor of safety $E(Q)$, whereas the upper one corresponds to the coefficient of variation of the stress V_Σ. From the value V_Σ on the upper horizontal axis, say $V_\Sigma = 0.14$, we draw a vertical downward line to meet the appropriate curve of constant V_{Σ_y}, say $V_{\Sigma_y} = 0.1$. From the point of intersection, denoted in the figure as (a), we draw a horizontal line that crosses the required level of reliability line. In the case illustrated in Fig. 8.7, the required reliability is taken to equal $r = 0.95$. The point of intersection is denoted (b). A vertical drop on the bottom horizontal axis will indicate the minimum value of the mean factor of safety $E(Q)$. In this particular case this minimum value equals about 4, finally, a straight line drawn from V_Σ and $E(Q)$ crosses the factor of safety line n_1 at the smallest required value n_1 denoted by (c). In this case $n_1 = 3.8$ approximately. My Dao-Thien and Massoud (1974) note that, "The monogram can be used for different practical purposes depending upon the known and unknown parameters."

8.6   Examples

1) Compare the calculated reliability of a machine beam to the predicted reliability of the beam given by the design nomogram. The strength distribution for this case has been determined to be given by a Weibull distribution where $E(\Sigma_y) = 74\,\text{MPa}$ and $V_{\Sigma_y} = 0.1318$. The mean stress is $E(\Sigma) = 55\,\text{MPa}$ and $V_\Sigma = 0.05$. Lipson and Narendra (1974) have calculated the exact reliability as 97 percent. We evaluate the central safety factor

$$n_1 = \frac{74}{55} = 1.35 \tag{8.82}$$

Using the above parameters, we find the predicted level of reliability from the design nomogram as follows:

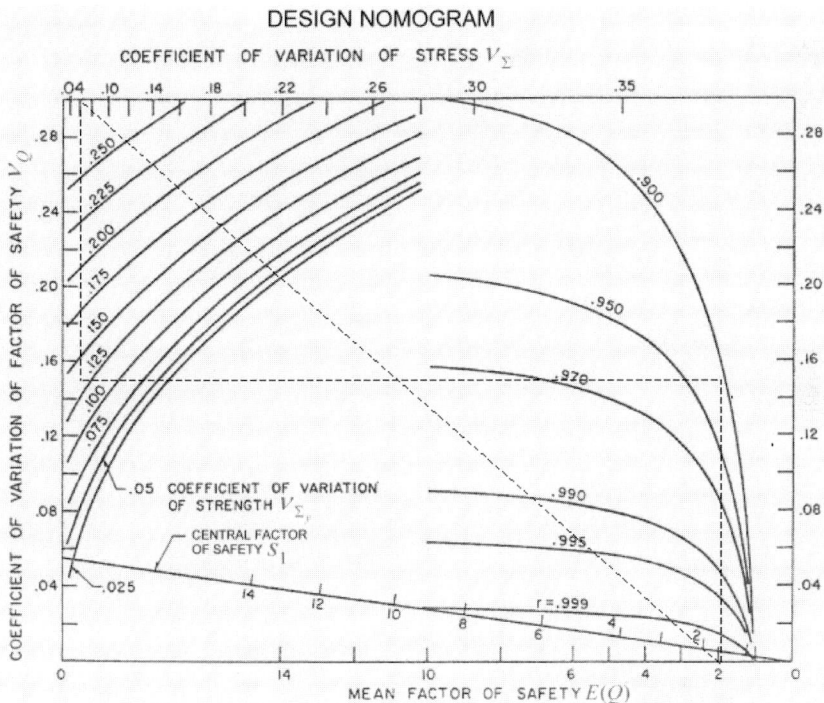

Fig. 8.8   Design nomogram: evaluation of examples

Therefore, it can be seen that the lower bound of reliability given by the design nomogram is approximately equal to 92%. This is the lower bound of the reliability. Indeed, the exact value 0.97 exceeds this lower bound value. The best use of the

172

nomogram is accomplished when we do not posses the information about the probability distributions of the stress and strength.

2) In the second example, both actual stress and yield strength have normal probability densities where $E(\Sigma_y) = 15,000\,\text{MPa}$, $v_{\Sigma_y} = 0.2$, $E(\Sigma) = 10,000\,\text{MPa}$ and $v_{\Sigma} = 0.1$. The exact reliability has been calculated as 94.305 percent (see Section 5.2). The central safety factor equals

$$n_1 = \frac{15,000}{10,000} = 1.5 \tag{8.83}$$

Using the above parameters, we find the predicted level of reliability from the design nomogram as illustrated below:

DESIGN NOMOGRAM

COEFFICIENT OF VARIATION OF STRESS V_Σ

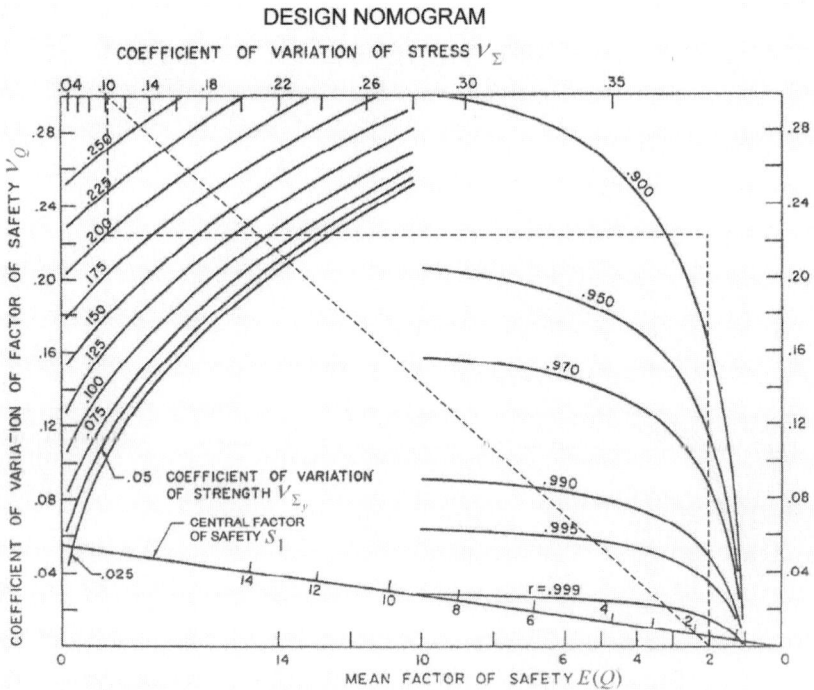

Fig. 8.9   Design nomogram: evaluation of examples

Therefore, it can be seen that the lower bound of reliability given by the design nomogram is approximately equal to 89%.

8.7   Conclusion: Other Bounds of Probability of Failure

As Melloy and Cavalier (1989) mention,

"Methods have been devised which determine the bounds of unreliability based upon the indpendent probabilities…, no distribution identification or knowledge of the means or variance is required. Unfortunately, empirical evidence demonstrates that existing methods can provide inaccurate bounds on the unreliability."

Kapur (1975) proposed a method for evaluating the bounds on the exact probability of failure which demand only information about the "interference" tail probabilities. This is done by partitioning the "interference region" into n stress and strength probability subintervals

$$\begin{cases} p_i = Prob\{a_{i-1} < \Sigma \le a_i\} \\ q_i = Prob\{a_{i-1} < \Sigma_y \le a_i\}, \end{cases} \quad \text{for } i = 1,2,..,n \tag{8.84}$$

where a_{i-1} and a_i define the endpoints of subinterval i. The probability of failure is then estimated as follows

$$P_f \approx \sum_{i=1}^{n} p_i \left[\sum_{k=1}^{i} q_k \right] \tag{8.85}$$

In practice, however, the exact probabilities cannot be obtained, therefore, the point estimates are bounded from below and above to form interval estimates for the individual and cumulative tail values:

$$\begin{cases} L_{p_i} \le p_i \le U_{p_i} \\ L_{q_i} \le q_i \le U_{q_i} \\ a_p \le \sum_{i=1}^{n} p_i \le b_p \\ a_q \le \sum_{i=1}^{n} q_i \le b_q \end{cases} \tag{8.86}$$

As Kapur (1975) notes, these interval estimates are based on "…experience and intimate knowledge of the operating environment." Park and Clark (1986) have demonstrated that the intervals generated by this method do not always bound the exact probability of failure. They suggested to use the modified variant of Eq. (8.85):

$$P_f \approx \sum_{i=1}^{n} p_i \left[\sum_{k=1}^{i-1} q_k \right] + 0.5 \sum_{i=1}^{n} p_i q_i \tag{8.87}$$

Melloy and Cavalier (1989) constructed a counterexample in which the method of Park and Clark (1986) yielded an interval [0.012797,0.013861] for the probability of failure, whereas its exact value constituted 0.010778. As is seen the probability of failure is overestimated by the method of Park and Clark (1986). The "cure" for this problem has been devised by Melloy and Cavalier (1989). The bounds for the probability of failure provided by the merhods of Kapur (1975), Park and Clark (1986) and Melloy and Cavalier(1989) are listed in Table 8.1.

Table 8.1. Bounds on the probability of failure for the three methods ($n = 10$)

Methods	Bounds on P_f
Exact	0.010778
Kapur (1975)	[0.019848 , 0.021498]
Park & Clark (1986)	[0.012797 , 0.013861]
Melloy & Cavalier (1989)	[0.005746 , 0.021498]

Chapter 9
Japanese Contributions to the Interrelating Safety Factor and Reliability

> "In structural reliability analysis, one must assume distribution functions for both loads and structural strengths, though the calculated structural reliability is not of good accuracy for lack of data of the distribution at levels which seldom take place, and furthermore it is very sensitive to the choice of the distribution functions."
>
> I. Konishi (1975)

> "The pace of accepting risk analysis has grown markedly in the last few years. New specifications for steel buildings, highway bridges, offshore platforms, etc. have been officially adopted and openly admit to applying risk analysis. (Actually, "reliability" analysis is usually the preferred term in the structural literature for obvious reasons…). Similar developments involving the acceptance of reliability-based specifications have occurred worldwide including Canada, Japan and the European Community. The major differences being the terminology and database rather than any disagreement in basic risk assessments."
>
> F.Moses (1998)

> "As always, our favorite subject is the calculation of structural reliability. To compute it, we must begin with probabilistic characterization of the actions against which we wish to design and of material behavior, move to the behavior and reliability of structural components and hence to structural reliability."
>
> E. Rosenblueth (1991)

As it was shown in Chapter 6 the probability of failure is a very sensitive to the exact shapes of the probability density function of the stress and strength, respectively. This gave rise to a non-probabilistic safety factor, proposed in the Chapter 6. Another, alternative approach is the use of the estimates of the safety factors that do not depend on particular shape of the probability densities of the stress and strength. Such approaches have been advanced by various investigators, notably by Mischke (1970), Dao-Thien and Massoud (1974) and Ichikawa (1983). We will be mostly following Ichikawa's (1983) and Reiser's (1985) works in exposition of this subject.

9.1 Introduction

Mischke (1970) employed the non-parametric, distribution-free approach using the Bienaymé inequality to derive the upper bound on the probability of failure at the fixed central safety factor. Dao-Thien and Massoud (1974) improved Mischke's results as follows :

$$P_f < \frac{n^2\left[\left(1+v_{\Sigma_y}^2\right)\left(1+3v_\Sigma^2\right)+k(k-2)\left(1+v_\Sigma^2\right)^2\right]}{\left[k\,n\left(1+v_\Sigma^2\right)-1\right]^2} \tag{9.1}$$

where the coefficient k is given by

$$k = \frac{n\left[\left(1+v_{\Sigma_y}^2\right)\left(1+3v_\Sigma^2\right)/\left(1+v_\Sigma^2\right)^2 -1\right]}{n\left(1+v_\Sigma^2\right)-1} \tag{9.2}$$

where, as in previous chapters, v_Σ and $v_{\Sigma y}$ are coefficients of variation of the actual stress and the yielding stress, respectively.

If the acceptable level p_f of the probability of failure P_f is specified, one can derive the following upper bound on the central factor of safety n:

$$n < \frac{1}{1+v_\Sigma^2 - \left\{\left(1-p_f\right)/p_f\left[\left(1+v_{\Sigma_y}^2\right)\left(1+3v_\Sigma^2\right)-\left(1+v_\Sigma^2\right)^2\right]\right\}^{1/2}} \tag{9.3}$$

Ichikawa (1983) derived an improved formula, which will be reproduced below.

9.2 Ichikawa's Formula

Consider the safety margins $M = \Sigma_y - \Sigma$. Then

$$P_f = Prob(M \le 0) \tag{9.4}$$

Naturally, one can write the following inequality

$$P_f < Prob(|M - \alpha| \ge \alpha) \tag{9.5}$$

where α is an arbitrary positive value. According to the Bienaymé inequality

$$Prob(|M - \alpha| \ge \alpha) \le \frac{E\left[(M-\alpha)^2\right]}{\alpha^2} \tag{9.6}$$

where $E(\cdot)$ denotes the mathematical expectation of $(\cdot)$. Bearing in mind Eq. (9.5) and (9.6) we arrive at

$$P_f < \frac{E\left[(M-\alpha)^2\right]}{\alpha^2} \tag{9.7}$$

Ichikawa (1983) reasoned as follows. We first note that

$$E\left[(M-\alpha)^2\right] = E(M^2) - 2\alpha E(M) + \alpha^2$$

and

$$(9.8)$$

$$E(M^2) = [E(M)]^2 + \sigma_M^2$$

Let us calculate the minimum value, with respect to the parameter α, of the ratio

$$\beta = \frac{E(M - \alpha)^2}{\alpha^2} \qquad (9.9)$$

We rewrite β as follows

$$\beta = \frac{E(M^2)}{\alpha^2} - 2\frac{E(M)}{\alpha} + 1 \qquad (9.10)$$

The requirement

$$\frac{d\beta}{d\alpha} = 0 \qquad (9.11)$$

yields

$$-2E(M^2)\alpha^{-1} + 2E(M) = 0 \qquad (9.12)$$

or

$$\alpha = \frac{E(M^2)}{E(M)} = \frac{[E(M)^2] + \sigma_M^2}{E(M)} \qquad (9.13)$$

The value attained by β at this value of α is

$$\beta = \frac{\sigma_M^2}{[E(M)]^2 + \sigma_M^2} \qquad (9.14)$$

Thus,

$$P_f < \frac{\sigma_M^2}{[E(M)]^2 + \sigma_M^2} \qquad (9.15)$$

If one assumes the mutual independence of Σ and Σ_y, it follows that

$$\sigma_M^2 = \sigma_\Sigma^2 + \sigma_{\Sigma_y}^2 \qquad (9.16)$$

where σ_Σ^2 and $\sigma_{\Sigma_y}^2$ are the variances of Σ and Σ_y, respectively. Substitution of Eq. (9.16) and of

$$E(M) = E(\Sigma_y) - E(\Sigma) \qquad (9.17)$$

into Eq. (9.15) yields

$$P_f < \frac{\sigma_\Sigma^2 + \sigma_{\Sigma_y}^2}{\sigma_\Sigma^2 + \sigma_{\Sigma_y}^2 + [E(\Sigma_y) - E(\Sigma)]^2} \qquad (9.18)$$

Ichikawa (1984) rewrote the expression (9.15) in terms of central safety factor n and the coefficients of variation v_Σ and v_{Σ_y};

$$P_f < \frac{n^2 v_{\Sigma_y}^2 + v_\Sigma^2}{(n-1)^2 + n^2 v_{\Sigma_y}^2 + v_\Sigma^2} \tag{9.19}$$

Eq. (9.16) yields an upper bound of the probability of failure for an given value of the central safety factor n. Solving Eq. (9.16) for n, we obtain the upper bound of n that or required to surpass a given probability failure P_f:

$$n < \frac{P_f + \left\{ P_f^2 - \left[\left(1 + v_{\Sigma_y}^2 \right) P_f - v_{\Sigma_y}^2 \right] \left[\left(1 + v_\Sigma^2 \right) P_f - v_\Sigma^2 \right] \right\}^{1/2}}{\left(1 + v_{\Sigma_y}^2 \right) P_f - v_{\Sigma_y}^2} \tag{9.20}$$

Ichikawa (1983) claims that the upper bounds of P_f and n determined by Eqs. (9.19) and (9.20) are considerably smaller than those from Eq. (9.1) and (9.3), respectively.

9.3 Reiser's Correction

Reiser (1985) noticed that an additional condition is required for Ichikawa's (1983) results to hold. Reiser notes that in the expression for the probability of failure

$$P_f = Prob(\Sigma_y \leq \Sigma) \tag{9.21}$$

if $E(\Sigma)$ tends to infinity with $E(\Sigma_y)$, while σ_{Σ_y} and σ_Σ are remaining fixed, P_f tends to unity for many different choices of distributions of Σ and Σ_y. Reiser (1985) notes:

"This is especially easy to see if it is assumed that Σ and Σ_y are normally distributed. However, as $E(\Sigma) \to \infty$, the right-hand side of the inequality (9.18) approaches zero. Thus a contradiction exist which calls into question the validity of Eq. (9.18)"

[The quote is adjusted to our notation and our sequence numbers of equations].

Reiser (1985) notes that α in Eq. (9.13) must be positive. Therefore, Eq. (9.13) is correct only for

$$E(M) \geq 0 \tag{9.22}$$

For the cases

$$E(M) < 0 \tag{9.23}$$

expression (9.9) for β approaches the minimum value of unity as $\alpha \to \infty$. Thus we conclude that

$$min\, E[(M - \alpha)^2]/\alpha^2 = \begin{cases} 1, & for\ E(M) < 0 \\ \dfrac{\sigma_M^2}{\sigma_M^2 + [E(M)]^2}, & for\ E(M) \geq 0 \end{cases} \tag{9.24}$$

Therefore, the equation (9.18) must be replaced by

$$P_f = \begin{cases} 1, \, if \; E(\Sigma_y) < E(\Sigma) \\ \dfrac{\sigma_\Sigma^2 + \sigma_{\Sigma_y}^2}{\sigma_\Sigma^2 + \sigma_{\Sigma_y}^2 + [E(\Sigma_y) - E(\Sigma)]^2}, \, if \; E(\Sigma_y) \geq E(\Sigma) \end{cases}$$ (9.25)

This formula will be referred to as Ichikawa-Reiser formula. About Reiser's (1985) correction Ichikawa (1986) noted:

"Reiser (1985) emphasized that Eq. (9.18) has a contradiction in that $P_f \to 0$ when $E(\Sigma) \to \infty$, with $E(\Sigma_y)$, σ_{Σ_y} and σ_Σ remaining fixed, and suggested the condition $E(\Sigma_y) > E(\Sigma_s)$ therefore needs to be added to Eq. (9.18). This conclusion was not added in (Ichikawa, 1983) as in practice the mean value of the stress is always made lower than that of the strength; hence, formal application ... to the unrealistic case, $E(\Sigma_y) < E(\Sigma)$ was not considered."

It is right, that one anticipates the central safety factor $n = E(\Sigma_y)/E(\Sigma)$ to be greater than unity. Still, this point does not appear to be obvious; hence, the clarification of Ichikawa's (1983) formula by Reiser (1985) appears to be extremely useful.

9.4   Another Set of Formulas by Ichikawa and Reiser

If one uses the truncated Taylor series approximation another expression for the bound can be obtained. Consider the distribution of the random variable

$$M = \ln \Sigma_y - \ln \Sigma$$ (9.26)

Then, in a manner perfectly analogous to the derivation of Eq. (9.25) we obtain

$$P_f = \begin{cases} 1, \qquad if \; E(\Sigma_y) < E(\Sigma) \\ \dfrac{\sigma_{\ln \Sigma_y}^2 + \sigma_{\ln \Sigma}^2}{\sigma_\Sigma^2 + \sigma_{\Sigma_y}^2 + [E(\ln \Sigma_y) - E(\ln \Sigma)]^2}, \quad if \; E(\Sigma_y) \geq E(\Sigma) \end{cases}$$ (9.27)

where $E(\ln \Sigma_y)$ and $E(\ln \Sigma)$ are the mean values of $\ln \Sigma_y$ and $\ln \Sigma$, respectively; $\sigma_{\ln \Sigma_y}^2$ and $\sigma_{\ln \Sigma}^2$ are the variances of $\ln \Sigma_y$ and $\ln \Sigma$, respectively. Use of the truncated Taylor series results in

$$E(\ln \Sigma_y) \approx \ln(\Sigma_y) + \frac{1}{2}\left(\frac{d^2 \ln \Sigma_y}{d\Sigma_y^2}\right)_{\Sigma_y = E(\Sigma_y)} \cdot \sigma_{\Sigma_y}^2 = \ln(\Sigma_y) - \frac{1}{2}[E(\Sigma_y)]^2$$

$$E(\ln \Sigma) \approx \ln(\Sigma) - \frac{1}{2}E[(\Sigma)]^2$$ (9.28)

$$\sigma_{\ln \Sigma_y}^2 = \left(\frac{d \ln \Sigma_y}{d\Sigma_y}\right)^2_{\Sigma_y = E(\Sigma_y)} \quad \sigma_{\Sigma_y}^2 = [E(\Sigma_y)]^2$$

$$\sigma_{\ln \Sigma}^2 \approx [E(\Sigma)]^2$$

Substitution of expression (9.28) into Eq. (9.27) yields

$$P_f = \begin{cases} 1, & \text{if} \quad E(\Sigma_y) < E(\Sigma) \\ \dfrac{[E(\Sigma_y)]^2 + [E(\Sigma)]^2}{[E(\Sigma_y)]^2 + [E(\Sigma)]^2 + \left\langle \ln n - 0.5\left\{[E(\ln \Sigma_y)]^2 - [E(\ln \Sigma)]^2\right\}\right\rangle^2}, & \text{if} \quad E(\Sigma_y) \geq E(\Sigma) \end{cases}$$

(9.29)

In case $E(\Sigma_y)$ is greater than or equal to $E(\Sigma)$ Eq. (9.28) allows to obtain the following upper bound for the central safety factor

$$n \leq \left\langle \exp\frac{[E(\Sigma_y)]^2 - [E(\Sigma)]^2}{2} + \left\{\frac{1-P_f}{P_f}[E(\Sigma_y)]^2 + [E(\Sigma)]^2\right\}^{\frac{1}{2}}\right\rangle$$

(9.30)

Careful examination of the Fig. 3, curves (a), (b), and (c) in the work by Ichikawa (1983) shows that the Reiser (1985) condition $E(\Sigma_y) \geq E(\Sigma)$ is violated. The Fig. 3d of Ichikawa's (1983) is reproduced here as Fig. 9.1.

9.5 Application of the Camp-Meidell Inequality

Let $E(M)$ be a positive quantity. Then the probability of failure

$$P_f = Prob(M \leq 0)$$

(9.31)

can be founded as follows

$$P_f < Prob[|M - E(M)| \geq E(M)]$$

(9.32)

The use of Camp-Meidell inequality yields

$$Prob[|M - E(M)| \geq E(M)] \leq \frac{4}{9}\frac{\sigma_M^2}{[E(M)]^2}$$

(9.33)

Hence,

$$P_f < \frac{4}{9}\frac{[E(\Sigma_y)]^2 + [E(\Sigma)]^2}{[E(\Sigma_y) - E(\Sigma)]^2}$$

(9.34)

or, in terms of the central safety factor n this inequality reads

$$P_f < \frac{4}{9}\frac{n^2[E(\Sigma_y)]^2 + [E(\Sigma)]^2}{(n-1)^2}$$

(9.35)

It is remarkable that the upper bound of P_f furnished by Eq. (9.35) is about half of that provided by Eq. (9.18). Ichikawa (1983) provides the following bound for the central safety factor

$$n \le \left\langle \frac{2.25P_f + \{2.25P_f [E(\Sigma_y)]^2 + [E(\Sigma)]^2 - [E(\Sigma_y)]^2 [E(\Sigma)]^2\}^{\frac{1}{2}}}{2.25P_f - [E(\Sigma_y)]^2} \right\rangle \tag{9.36}$$

Likewise, if the distribution of $M = ln\Sigma_y - ln\Sigma$ is continuos and unimodal, and if $E(M)$ coalesces with the mode of the random variable M, the following bound was derived by Ichikawa (1983) by using the Camp-Meidell inequality if the Reiser (1985) condition $E(\Sigma_y) \ge E(\Sigma)$ is met:

$$P_f < \frac{4}{9} \frac{\sigma_{ln\Sigma_y}^2 + \sigma_{ln\Sigma}^2}{[E(ln\Sigma_y) - E(ln\Sigma)]^2} \tag{9.37}$$

or, in other words,

$$P_f < \frac{4}{9} \frac{[E(\Sigma_y) - E(\Sigma)]^2}{\left\langle ln\, n - 0.5\{[E(\Sigma_y)]^2 - [E(\Sigma)]^2\} \right\rangle} \tag{9.38}$$

$$n < \exp\left\langle \frac{[E(\Sigma_y)]^2 - [E(\Sigma)]^2}{2} + \left\{ \frac{[E(\Sigma_y)]^2 - [E(\Sigma)]^2}{2.25P_f} \right\}^{\frac{1}{2}} \right\rangle \tag{9.39}$$

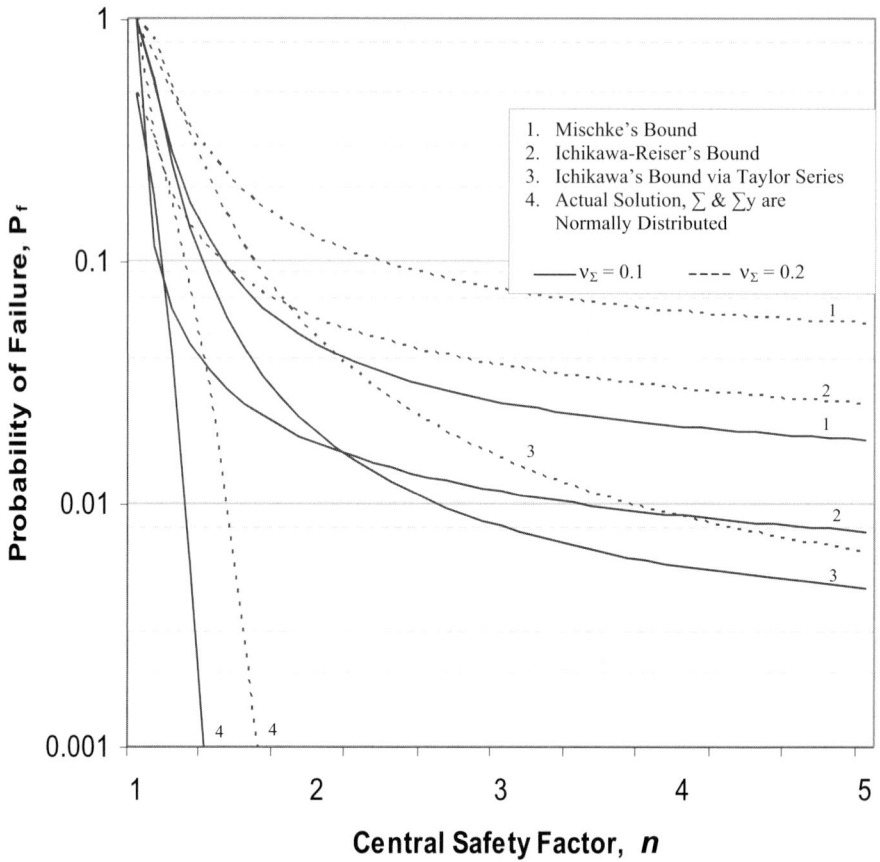

Fig. 9.1 Comparison of the Ichikawa-Reiser bounds with those of Mischke
$$(v_{\Sigma y} = 0.05)$$

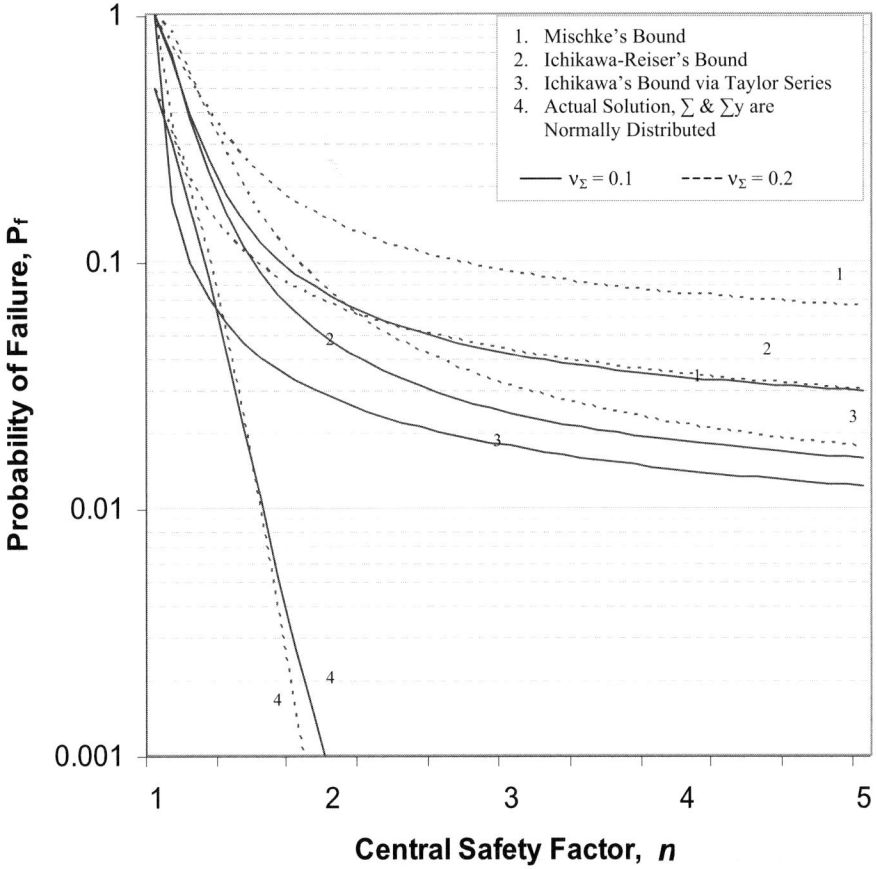

Fig. 9.2 Comparison of the Ichikawa-Reiser bounds with those of Mischke
$$(\nu_{\Sigma y} = 0.10)$$

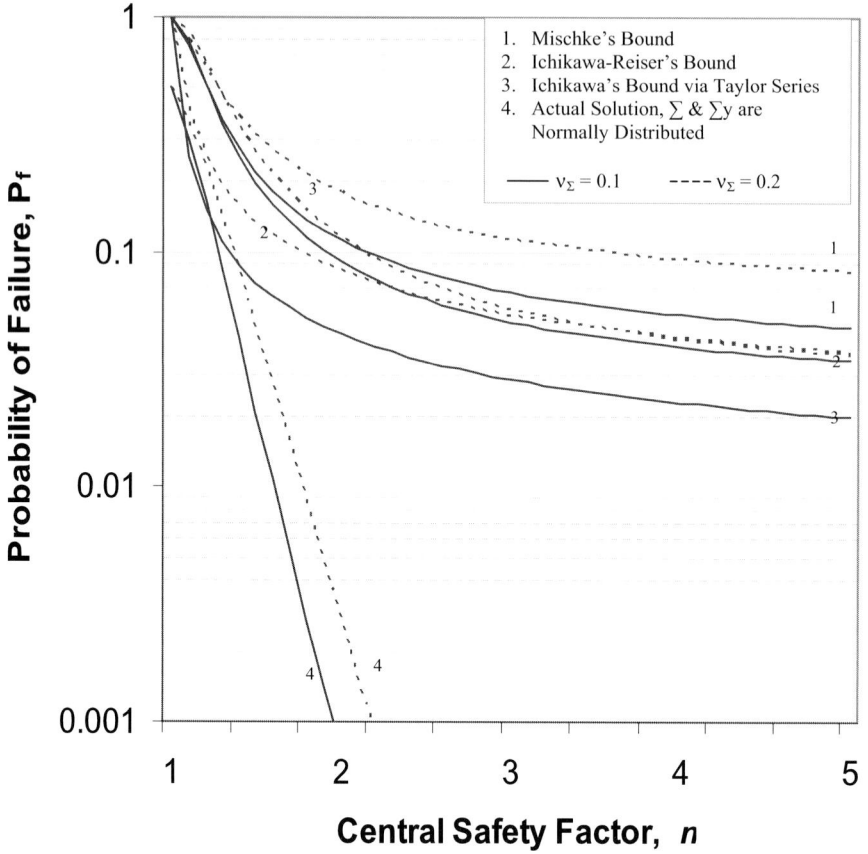

Fig. 9.3 Comparison of the Ichikawa-Reiser bounds with those of Mischke
$(\nu_{\Sigma y} = 0.15)$

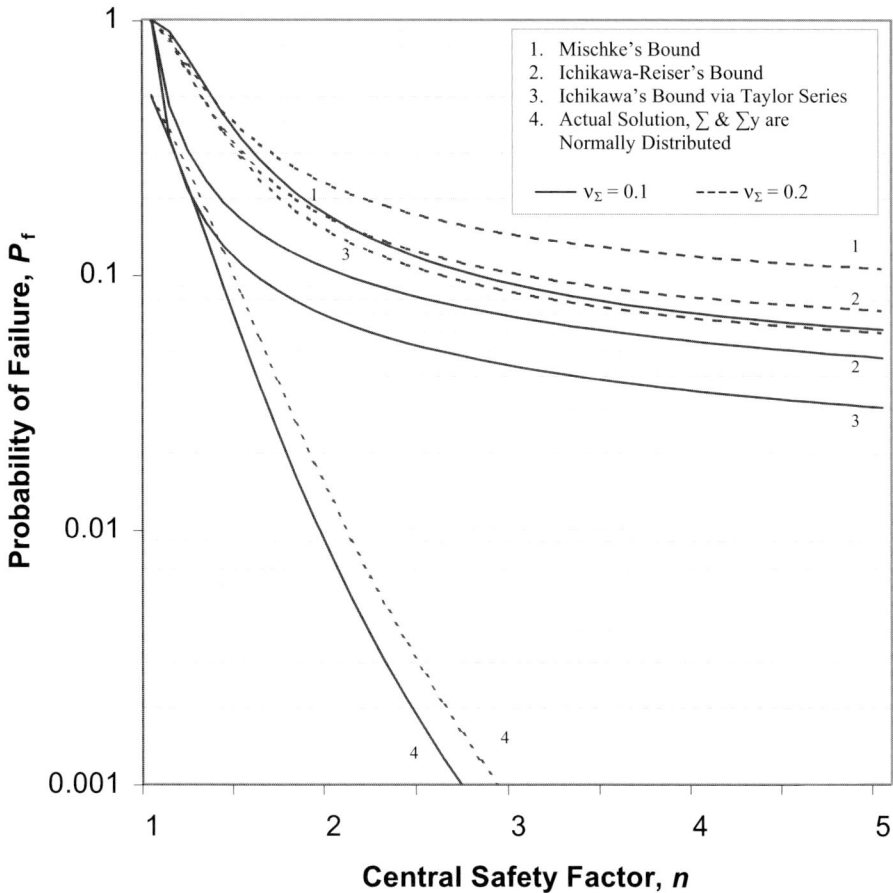

Fig. 9.4 Comparison of the Ichikawa-Reiser bounds with those of Mischke
($v_{\Sigma y} = 0.20$)

9.6   Series Representation of the Probability Density Functions

Harold Cramér (1970) discusses a method for finding an accurate approximation of a continuous probability distribution function expressed through the normal distribution function, since the latter easy to manipulate analytically. We use the following representation of the arbitrary probability density function via the normal one:

$$f_X(x) = \phi(x) + r(x) \qquad (9.40)$$

where $\phi(x)$ is a normal probability density function with zero mean and unit variance, $r(x)$ is the remainder term. Utilizing a method involving orthogonal polynomials we can replace $f(x)$ as

$$f_X(x) = c_0\phi(x) + \frac{c_1}{1!}\phi'(x) + \frac{c_2}{2!}\phi''(x) + \frac{c_3}{3!}\phi'''(x) + \frac{c_4}{4!}\phi^{IV}(x) + \dots \qquad (9.41)$$

where the αth derivative is connected with $\phi(x)$ as follows:

$$\phi^{(\alpha)}(x) = (-1)^\alpha H_\alpha(x)\phi(x), \qquad (9.42)$$

$H_\alpha(x)$ being the Hermite polynomial

$$H_\alpha(x) = (-1)^\alpha e^{\frac{x^2}{2}} \frac{d^\alpha}{dx^\alpha}\left(e^{-\frac{x^2}{2}}\right) \qquad (9.43)$$

The constant coefficients c_α in Eq. (9.2) are defined as follows:

$$c_\alpha = (-1)^\alpha \int_{-\infty}^{\infty} H_\alpha(x)f(x)dx \qquad (9.44)$$

They equal:

$$c_0 = 1, \quad c_1 = 1, \quad c_2 = 1$$

$$c_3 = -\frac{\mu_3}{\sigma^3}$$

$$c_4 = \frac{\mu_4}{\sigma^4} - 3 \qquad (9.45)$$

$$c_5 = -\frac{\mu_5}{\sigma^5} + 10\frac{\mu_3}{\sigma^3}$$

$$c_6 = \frac{\mu_6}{\sigma^6} - 15\frac{\mu_4}{\sigma^4} + 30$$

representation of $f_X(x)$. When it is required to include a further term in the expansion, it is often recommended in the literature that a following term containing the fourth derivative $\phi^{(4)}(x)$ be considered. The theoretically correct rule is, however, to let the following term contain both the fourth and the sixth derivatives, thus arriving at the expansion

$$f_X(x) = \phi(x) + \frac{\gamma_1}{3!}\phi^{(3)}(x) + \frac{\gamma_2}{4!}\phi^{(4)}(x) + \frac{10\gamma_1^2}{6!}\phi^{(6)}(x) + \dots \qquad (9.46)$$

γ_2 being the measure of excess [γ_1 being the measure of skewness]. The reason for preferring this expansion is that it can be shown that the coefficients of the terms in

$\phi^{(4)}(x)$ and $\phi^{(6)}(x)$ are, under general conditions, of the same order of magnitude for large *n*. In practice it is usually not advisable to go beyond this expansion. By including the additional terms containing the derivatives, the symmetry of the normal function $\phi(x)$ is destroyed."

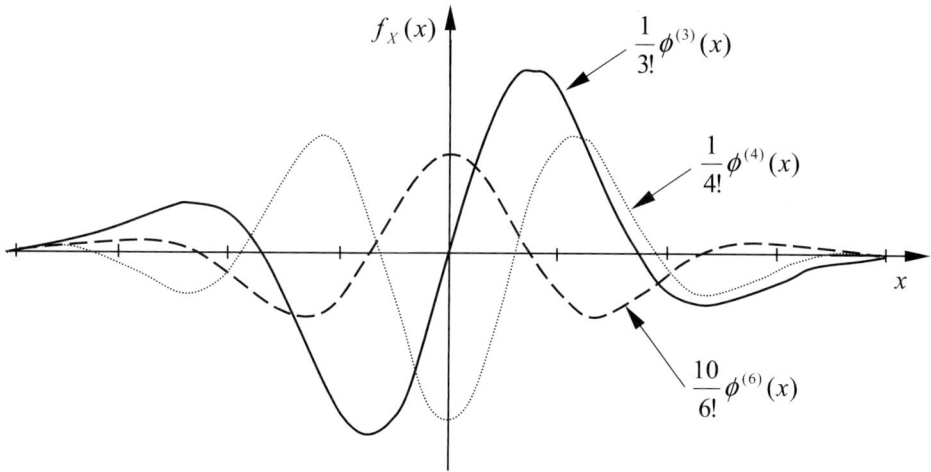

Fig. 9.5 Derivatives of the normal density function $\phi(x)$

At this point, the coefficients can be expressed in terms of the central moments μ_α and the standard deviation σ, arriving at the series of Francis Ysidro Edgeworth (1845-1926):

$$f_X(x) = \phi(x) - \frac{1}{3!}\frac{\mu_3}{\sigma^3}\phi^{(3)}(x) + \frac{1}{4!}\left(\frac{\mu_4}{\sigma^4} - 3\right)\phi^{(4)}(x) + \frac{10}{6!}\left(\frac{\mu_3}{\sigma^3}\right)^2\phi^{(6)}(x) + ...$$

$$(9.47)$$

Murotsu et al (1978) utilize Cramer's (1946) representation for deriving the expression for the probability of failure and the safety factor.

9.7   Use of the Edgeworth Series by Murotsu et al

This section follows closely the study by Murotsu et al (1978). Traditional methods for deriving the relationship between the safety factor and the failure probability, P_f is given by:

$$P_f = F_{M_s}\left(-\frac{E(M)}{\sigma_M}\right)$$

(9.48)

where

$$-\frac{E(M)}{\sigma_M} = \frac{1-n}{\sqrt{(s\,v_{\Sigma_y})^2 + v_{\Sigma}^2}}$$

(9.49)

and n is the central safety factor; P_f is the probability of failure; v_{Σ_y} is the coefficient of variation for Σy (the capacity of the structure - e.g., yield stress) and v_{Σ} is the coefficient of variation for Σ (the applied stress).

Note that $F_{Ms}(x)$ is the cumulative distribution function of the standardized safety margin. Murotsu et al (1978) state, "it is difficult to formulate explicitly $F_{Ms}(.)$ except the special cases where both Σ_y and Σ are distributed normally or log-normally [Note that explicit expressions are available for many other cases as shown in Chapter 5 - I.E.]. Consequently, the exact relationship between central factor of safety and failure probability can not be evaluated. Numerical integration and Monte Carlo simulation have been applied to evaluate the relationship between P_f and n for general combinations of probability distributions. However, they are not so efficient with respect to their accuracies and processing times... Therefore, it is difficult to evaluate the correspondence of safety factor to reliability larger than 0.99."

The cumulative distribution function for M_s treated as a random variable, derived from the expansion as expressed by Cramér (1946) is

$$F_{M_s}(x) = \Phi(x) - \frac{\beta_{M1}}{3!}\Phi^{(3)}(x) + \frac{\beta_{M2}-3}{4!}\Phi^{(4)}(x) + \frac{10\beta_{M1}^2}{6!}\Phi^{(6)}(x)$$

(9.50)

where β_{M1} is the coefficient of skewness and equals γ_1, and β_{M2} is the coefficient of kurtosis and equals $\gamma_2 + 3$. These coefficients read under the assumption of statistical independence between the actual stress and the yield stress:

$$\beta_{M1} = \frac{\beta_{\Sigma_y 1} - \beta_{\Sigma 1}(v_{\Sigma}^3/n^3 v_{\Sigma_y}^3)}{\{1+(v_{\Sigma}^2/n^2 v_{\Sigma_y}^2)\}^{3/2}}$$

(9.51)

and

$$\beta_{M2} = \frac{\beta_{\Sigma_y 2} + 6(v_{\Sigma}^2/n^2 v_{\Sigma_y}^2) + \beta_{\Sigma 2}(v_{\Sigma}^4/n^4 v_{\Sigma_y}^4)}{\{1+(v_{\Sigma}^2/n^2 v_{\Sigma_y}^2)\}^2}$$

(9.52)

where $\beta_{(.)1}$ is the coefficient of skewness of (.) and $\beta_{(.)2}$ is the coefficient of kurtosis of (.).

Substituting these values into the P_f formula the following relation evolves:

$$P_f = F_{M_s}\left(-\frac{1}{v_M}\right) = \Phi\left(-\frac{1}{v_M}\right) - \left[\frac{\beta_{M1}}{6}\left\{\left(-\frac{1}{v_M}\right)^2 - 1\right\} + \frac{\beta_{M2}-3}{24}\left(-\frac{1}{v_M}\right)\left\{\left(-\frac{1}{v_M}\right)^2 - 3\right\}\right.$$

$$\left. + \frac{\beta_{M1}^2}{72}\left(-\frac{1}{v_M}\right)\left\{\left(-\frac{1}{v_M}\right)^4 - 10\left(-\frac{1}{v_M}\right)^2 + 15\right\}\right]\Phi\left(-\frac{1}{v_M}\right)$$

$$(9.53)$$

Murotsu et al (1978) also explore the region where the probability density function "approximated by Edgeworth's series expansion gives negative values for the combinations of skewness and kurtosis. The shaded area in Fig 9.6 indicates the inapplicable region thus determined. The numbers in parentheses on the boundary designate the maximum values of $(-1/v_M)$ which yield negative failure probability in the region examined." The two following situations evolve to rectify the above observation:
(a) If $\beta_{M1} < 0$ and $\beta_{M2} > 3$, then

$$P_f = F_{M_s}\left(-\frac{1}{v_M}\right)$$

$$(9.54)$$

for all values of $-1/v_M$

If $\beta_{M1} > 0$ and $\beta_{M2} < 3$, then

$$P_f = \begin{cases} F_{M_s}\left(-\frac{1}{v_M}\right), & for \ -\frac{1}{v_M} < -1.5 \\ \dfrac{F_{M_s}(-1.5)\Phi\left(-\frac{1}{v_M}\right)}{\Phi(-1.5)}, & for \ -\frac{1}{v_M} < -1.5 \end{cases}$$

$$(9.55)$$

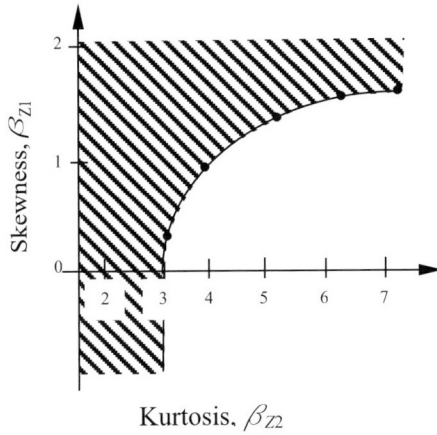

Fig. 9.6 Shaded area represents a region in which the approximation of the probability density function by the Egdeworth's series expansion is inapplicable.

From here a comparison between the actual numerical integration of the various distribution functions and the proposed method as described by Eq. (9.55) above. Some sample distributions are evaluated. In the following set of figures a solid line is used to represent the relation for numerical integration and a dotted line for the method by Murotsu et al (1978).

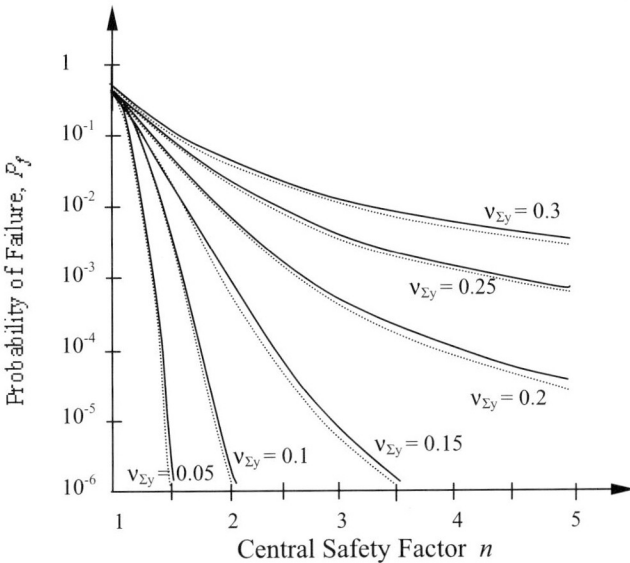

Fig. 9.7 Relation between P_f and n evaluated with Σ_y having a Normal Distribution and Σ having a Weibull distribution with v_Σ held constant at 0.1

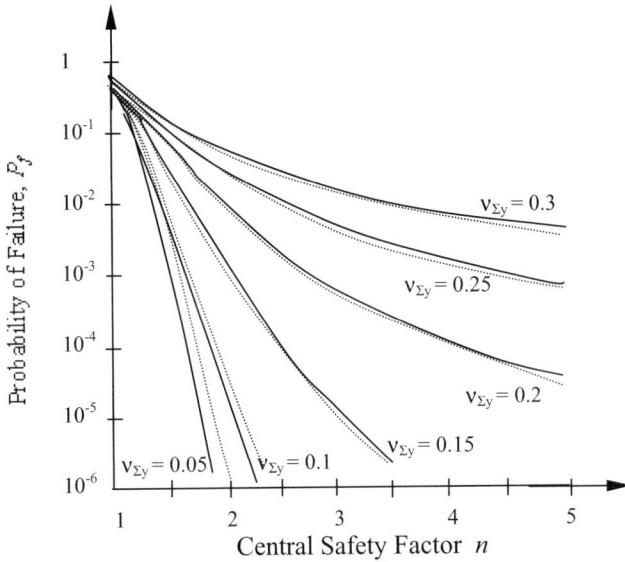

Fig. 9.8 Relation between P_f and n evaluated with Σ_y having a Normal Distribution and Σ following an extremal distribution with ν_Σ held constant at 0.1

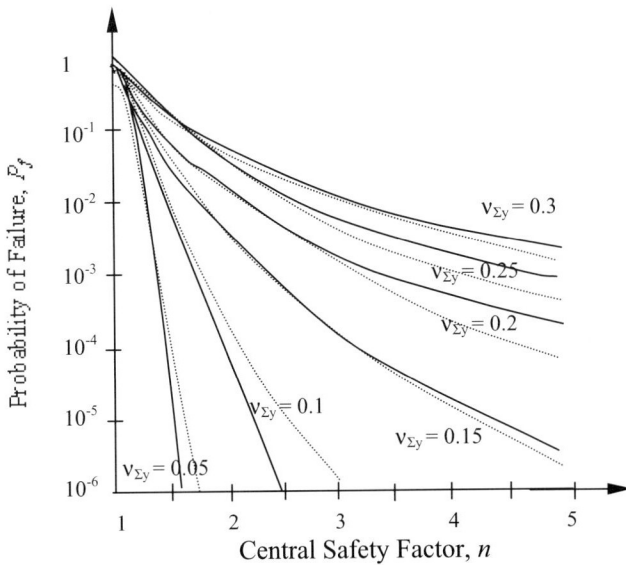

Fig. 9.9 Relation between P_f and n evaluated with Σ_y having a Weibull Distribution and having a Weibull distribution with ν_Σ held constant at 0.1

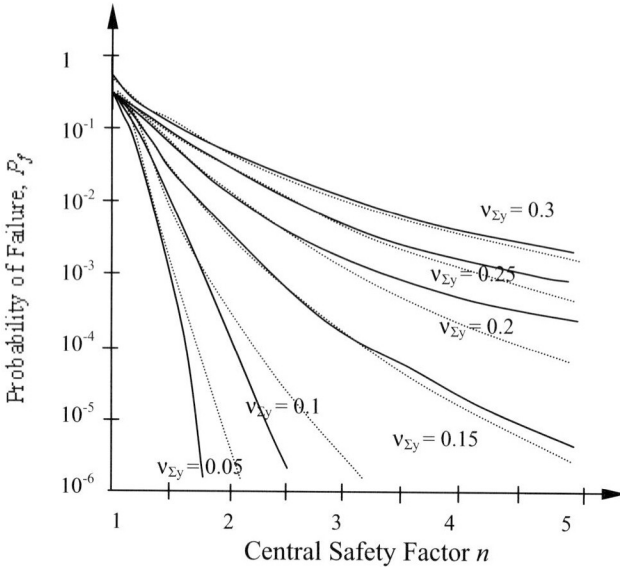

Fig. 9.10 Relation between P_f and n evaluated with Σ_y having a Weibull Distribution and Σ following an extrema₁ distribution with v_Σ held constant at 0.1

The method by Murotsu et al (1978) is practically identical to the numerical integration. There is significant enough agreement with the understanding that at P_f values smaller than 10^{-3}, their method is on the "safe side" of the numerical integration for large values of v_Σ. On the other hand for small values of v_Σ the values for the Murotsu et al (1978) method cross out of the safe side for P_f values smaller than 10^{-3}.

According to these results, the probability of failure can be considered very sensitive to the combination of distributions for the applied load and the yield stress. Therefore, Murotsu et al (1978) conclude that there must always be "a rational selection of underlying probability distributions" in the probabilistic design of structures.

9.8   Hoshiya's Distinction of Seemingly Equivalent Designs

The following question was raised by Hoshiya (2000):

"(1) Which one should be employed if two alternatives have a same level of reliability, but have noticeably different degrees of uncertainty?"

He considered the following example of basic reliability problem in which the performance function was given as follows:

$$Z = \Sigma_y - \Sigma \qquad (9.56)$$

For simiplicity, both Σ_y and Σ were assumed to be normally distributed random variables that were mutually independent. The stress Σ has a mean value $E(\Sigma)=2$ and the standard deviation $\sigma_\Sigma = 1.5$. Units are omitted purposely since they are not pertinent to the discussion. We have two designs associated with two different normal distributions for the yield stress, Σ_y:

$$\text{Design A: } E(\Sigma_y) = 8, \sigma_{\Sigma y} = 0 \qquad (9.57)$$
$$\text{Design B: } E(\Sigma_y) = 12, \sigma_{\Sigma y} = 2$$

These two designs possess the same probability of failure.

Hoshiya (2000) evaluated the central safety factor $n=E(\Sigma_y)/E(\Sigma)$ as well as the characteristic safety factor $\sigma_y*/\sigma*$, where the probability that Σ exceeds the value $\sigma*$ equals 0.05. In addition, he evaluated the information entropy

$$H = - \int_{-\infty}^{\infty} f_z(z) \ln f_z(z) dz \qquad (9.58)$$

of the performance function Z; $f_z(z)$ denotes the probability density of the function Z. Obviously, if the standard deviation of Z increases, so does the entropy, while for the deterministic case of Z, the entropy vanishes identically. Naturally, the greater entropy is associated with greater uncertainty in the design. Table 9.1 lists the characteristics for the designs A and B.

Table 9.1: Design Characteristics

Characteristics	Design A	Design B	Evaluation
P_f	3.17×10^{-5}	3.17×10^{-5}	$A \equiv B$
Central FS	4	6	B preferable
Characteristic FS	1.79	1.96	B preferable
H	1.8244	2.3352	A preferable
Total cost	$200	$200	$A \equiv B$

Following conclusions are deduced from the Table 9.1:

(1) In terms of the associated probabilities of failure, designs A and B are equivalent.
(2) If one includes into the characterization the central safety factor, design B appears to be better.
(3) If one includes into characterization the entropy, design A appears to be better, as associated with less uncertainty.
(4) Let us include in the characterization the cost. Since design A has less material deviation and, thus, the material used must be more expensive and the unit price is higher, resulting in less volume of material than for design B. On the other hand, design B may utilize relatively inexpensive material, with lower unit price, and hence more volume of the material. Therefore, designs A and B may turn out to be nearly equivalent costwise.

Hoshiya (2000) concludes:

"....[it] is better to choose design *A* based on the degree of uncertainty, despite number of the same reliability. This decision could not have been made if only the reliability level were to be given to engineers."

9.9   Contribution by Konishi et al: Proof Loads

In their paper, Konishi et al (1975) mention: "In structural reliability analysis, one must assume distribution functions for both loads and structural strength, though the calculated structural reliability is not of good accuracy for lack of data of the distributional level which seldom take place, and furthermore it is very sensitive to the choice of the distribution function." This difficulty can be overcome if proof tests are performed for the structural elements. This means that the members with strength below some level are eliminated from the statistical population. Besides engineering benefits, there are also analytical ones. Since proof testing improves reliability the problem of sensitivity must be "solved to some extent" (Konishi et al, 1975, p. 92).

Konishi, Kitagawa and Katsuragi (1975) studied the effect of proof load test a structural reliability. They arrived at the following conclusions:

1.   "In the case of no proof load test, the distribution function is remarkably influenced on the probability of failure of a simple tension bar and the large the coefficient of variation of resistance, the stronger the tendency."

2.   "When the proof load test is performed, the distributional sensitivity of the reliability of a simple tension bar is considerably improved if the coefficient of variation of loads is less than 0.20."

3   "The reliability is, however, sensitive to distribution functions if the coefficient of variation of loads is more than 0.3."

Shinozuka and Yang (1969) noted the advantage of performing the proof load as follows:

" The test can improve not only the reliability value itself but also the statistical confidence in such a reliability estimate. This is because the proof load test eliminates structures with strength less than the proof load. In other words, the structure that passes the proof load test belongs to the subset, having the strength higher than the proof load, of the original population. Therefore, it is obvious that the reliability of a structure chosen from this subset is higher that of a structure chosen from the original population. Furthermore, the proof-load test truncates the distribution function of strength at the proof load, hence alleviating the analytical difficulty of verifying the validity of a fitted distribution function at the lower tail portion where data are usually non-existent."

Other definitive words on proof-loading are those by Barnett and Hermann (1965),Curnick (1970),Tiffany (1970),Evans and Widerhorn (1974),Yang (1976), Fujino and Lind (1977),Heller, Schmidt and Deminghoff (1984), Heller, Thangjitham and Wall (1986), Thangjitham and Heller (1987), Hong and Lind (1991), Yang et al (1993), Yang (1994).

The finding by Konishi et al (1975) about less sensitivity of the reliability in some ranges of the coefficient of variations was a tremendously *good news* for the probabilistic approach to design. This was due to the fact that some researchers question the very validity of the probabilistic design because of the "distributional sensitivity." Proof–test is good for the structural performance; it increases the reliability and, for moderate coefficient of variation, decreases the sensitivity, thus increasing our confidence in probabilistic design itself. This remarkable finding by Konishi et al (1975) reinforced another remarkable observation, by Shinozuka and Yang (1969), that "the significant improvement of statistical confidence in the reliability estimate can be achieved on the basis of proof–load test."

9.10   Concluding Remarks

In words of Ichikawa (1984),

"reliability demonstration tests may be classified into two cases; the distribution specified case and the distribution free case. In the former, a distribution model of the time to failure such as Weibull distribution is assumed, whereas it is not assumed in the latter case."

Naturally, the knowledge of the probability distributions of stress and strength is often lacking. This justifies the use of the inequalities by Ichikawa (1983) and Reiser (1985). It appears that the information entropy is a powerful concept to better distinguish the seemingly equivalent designs. Other fundamental contributions from Japan, in humble opinion of this writer, include development of the "stochastic finite element" methology by Nakagiri and Hisada (1965), as well as works by Kanda (1996) in various branches of introducing probabilistic methods into practice, to name just a few.

The Japanese school of structural reliability, pioneered apparently by Professor Konishi in the Sixties, continues to be an extremely vibrant one.

Chapter 10
Epilogue

"We all believe that, except perhaps at the atomic level the physical world is governed by deterministic laws. Given full information about the loading conditions the material properties, and the mode of construction of a structure, we should be able, given sufficient time, to envolve a perfect design, where the resistance of every member is exactly matched to the load imposed on it. However such perfect design is not attainable, because of the imperfection of our knowledge...

The classical way of dealing with this problem has been through the use of the factor of safety. But in recent years the need for a more precise and rational approach has been felt. Just as in the other technological fields, it has become established that the best way of dealing with situations involving uncertainty is through these of probability theory...

While the complete rationality of the above-mentioned method is not in doubt, it is too remote from accepted day-to-day practice that it will take many years until it become established as the standard method of design."

A.M. Hasofer (1970)

"It is true that the adaptation of a probabilistic approach is in any case a progress."

G. Grandori (1991)

"The conventional design methods based on the factor of safety are not rational in the sense that the some factor of safety might imply different values of reliability in different situations."

S.S. Rao (1992)

The natural question arises: "Probabilistic methods have been around quite for some time. When will they be introduced into design?" In their careful review Roësset and Yao (2002) write:

"Risk analyses have started to become standard requirements in these fields [earthquake engineering; nuclear engineering; offshore engineering].Yet the introduction in practice of probability concepts has been slow. Even when the probability-based load and resistance factors design (LRFD) specifications were introduced in design codes, they were used without explicit mention of probabilities. The load and resistance factors were selected through calibration, in order to obtain results similar to those of the working stress design and past experience, rather than on the basis of existing uncertainties."

A colleague of this writer, when discussing this state of affairs during the presentation of his lecture, quoted Max Planck:

"A new scientific truth does not triumph by convincing its opponents and making them see the light, but rather because its opponents finally die, and a new generation grown up that is familiar with it."

The presenter, who is an extremely active contributor to probabilistic mechanics, and co-organizer of numerous regional and international conferences on this topic, complained that his colleagues in the Mechanical and Civil Engineering Departments are quite reluctant to utilize probabilistic methods. He found Plank's above quote quite consoling. Still, to adopt Max Planck's aggressive "resolution" of the above question appears to be very pessimistic. This is so, especially in the light of the comment made by A.L. Mackay in relation to the above quote of Max Planck:

"How can we have any new ideas of fresh on hook when 90 percent of all the scientists who have ever lived have still not died?"

If Max Planck were not a famous physicist one would possibly take his quote as a humorous remark. Who knows, perhaps this is how he meant it, with twinkle in his eyes? Still we would like to be sensitive to his remark and perhaps modify it in a manner to state that new scientific theories need time to be fully absorbed and appreciated. Naturally, those who propagate a new paradigm ought to present it in a simple manner, rather than in the "esoteric" contexts.

It is instructive to quote here Schuëller (1993):
"The sophistication of mechanical modeling of structures has progressed considerably, particularly within the last three decades. This was mainly due to the development of the Finite Element method (FEM), which, by utilizing high-speed computers, proved to be computationally most efficient. This goes along with an equally dramatic development of material science which provided significant additional insight in material properties."
According to Levi (1979)

"Structural safety, that is to say, the discipline which deals with these problems [probabilistic principles and methods in the design and checking of structures], albeit a comparatively young science, has....already proved its worth."

The reader may inquire, if these developments prompted acceptance of the probabilistic methods. Schuëller (1993) writes:

"It is interesting to observe that, despite of these developments, analysis and design procedures, by and large, still followed the traditional lines, ie. selecting from the spectra of values for each parameter a particular value, e.g. "maximum" value for loads, "minimum" or mean values for material properties and mainly empirically based safety factors. In other words, the degree of accuracy between structural, i.e., stress analysis on one had and load, material and safety analysis on the other still diverge significantly, as it was already observed.... by Freudenthal (1964)."

198

Indeed, Freudenthal (1964) writes (free translation from the German):

> "For years, more or less unconnected teams of engineers have been making serious attempts to tackle the problem of structure safety, and it is of interest to note that their success – whether they are working in Spain, England, the USSR, Sweden or the U.S. – is not very encouraging. Construction engineers are still in general convinced that intuition and conventional prescriptions suffice for designing safe structures. And, it should be admitted that the conventional evaluation method seems adequate so long as oversizing had only financial consequences and merely makes for excessive material cost. When, however, the structure is to be designed for maximum performance capacity, and excessive material consumption impairs this capacity, an economical design approach becomes an absolute necessity.
>
> For this reason, the intuitive approach has broken down first of all in aircraft construction, and the concept of design for a specified failure probability was adopted for the first time in the aviation industry. In the building industry, where oversizing means only waste of money and large projects are usually financed by loans extending over several generations – this waste is not regarded as a serious drawback, all the more so as increased material consumption makes for increased labor employment, and moreover (especially in the U.S.) for reduced cost through standardization of the construction elements. In the aviation industry, by constrast, oversizing affects the operational capacity of the aircraft, and certain capacities are only attainable at minimized weight. In this case, it is thus essential to understand that design not based on an admissible failure probability is bound to hamper progress in aircraft construction."

It is instructive to quote here Lind,

> "It seems that every little attention was paid to reliability in the study of mechanics until recently. For whatever reason, the mainstream of mechanics has been concerned with deterministic problems, of the form: Given exactly a system and some input actions, determine the response…Random aspects were neglected in the study of mechanics. They were left for the engineer and builder to sort out."

And engineers responded to this challenge very ingeniously: They introduced the safety factor. What is the future holding for the non-deterministic mechanics? The answer is contained in the quote by Tichý (1991):

> "It can be said that a new profession, that of the Reliability Engineer, is just emerging at present, and that, in the future, reliability engineers will act as advisors to designers, contractors, and public authorities in most major construction projects, as well as in the everyday construction problems in the creative community."

It is hoped that you, respected reader, will participate in this fascinating endeavor.

Carter (1997, p. 194) stresses, as if resonating with the above discussion:

"But prejudices die hard! I can see that many older designers could wish to stick to the empirical factors with which they have so much experience. It is very understandable. I can see as one tries to represent each of the inputs statistically, the imponderables become greater and greater. I have no doubt that all materials properties could be evaluated statistically."

How the transition ought to proceed if adopted by the engineering community? Carter (1997, p.195) provides some partial answers to this inquiry:

"The final issue, so far as all is concerned, is the transparency of the design methodology based on statistically defined input data. Every feature of the design has to be separately and quantitatively defined. All the data can, ideally, be checked by test: even if it can prohibitive in time and cost actually to do so. Of more importance, all those input data can be and should be agreed by all involved, and can be subsequently audited; and given those input data, the outcome is independent of any subjective manipulation."

Moreover, he sums up as follows,

"I thus come down firmly on the side of a design methodology based on statistically defined input data; I hope that the reader agrees. The factor of safety has served us well over a century, and among others I should be sorry to lose a familial tool, but it is fundamentally flowered and must go – eventually.

For no one could – or should – dream of changing from on method to another overnight. The old and new methodologies must be used together until a permanent change is, or is not, justified. "

The situation described by Cornell (1981) two decades ago apparently has not changed

"Finally, there is relatively little 'selling' of the area and therefore very few disappointed 'buyers'; both merchants and consumers have an informed perception."

At the same time the observation made by Cornell (1969) earlier, still holds:

"The fundamental role of probability theory in safety and performance analysis is widely recognized in all branches of engineering. Despite more than 20 years of recognition in structural engineering literature, however, probabilistic methods have not yet been explicitly adopted as a basis for United States codes of standard practice."

Bolotin (1961) warns:

"There is a wide range of problems of structural mechanics in which the use of statistical methods may only play the role of auxiliary methods of analysis. Here the statistical and deterministic methods could successfully co-exist, complementing each other…the overestimation of the role of statistical methods can only be harmful."

Some researchers get carried away while "selling" the probabilistic method in mechanics. One of them even associates it with Heisenberg's uncertainty principle: "Until recently, many of the implications due to the probabilistic nature of engineering phenomena were not fully realized, even though from early in the

200

20th century, with the announcement of Heisenberg's uncertainty principle, the whole of science (including engineering) was recognized as ultimately based philosophically on the concepts of experimental probability."

Cornell (1969) is also warning against exaggerated optimism:

"On a more esoteric level, it is unreasonable to believe that a complete, strictly statistical probabilistic analysis of structural safety will ever be feasible for standard structures. The available data will always be too limited to provide wholly reliable established of failure probabilities of the order of magnitude necessary far adequate public safety. The sample sizes required are in multimillions. Extrapolation of mathematical models of frequency distributions will always be a part of these analyses…"

There are, then, sound reasons why the suggested adoption of probabilistic safety analysis has not been implemented in standard structural engineering practice. Nonetheless, the fundamental arguments for statistical basis remain. It still promises to provide a more rational, quantitative representation of engineering design."

In the Louvre Museum in Paris, France, the famous and a must-see painting "Mona Lisa" by Leonardo da Vinci is allowed to the viewed from a distance providing a fence for its security. Not unlike this, engineers treat every item or a structure with almost caution by using the fence around its safety. Up to now this was accomplished with great degree of success by ingenious introduction of safety factors. But if seems the time is ripe to move from pure empiricism to science. As is shown in this book, probabilistic or convex analyses could serve as more useful scientific frameworks for allocating safety factors.

It appears pertinent to understand, that, as Thoft-Christensen (1980) emphasized, "The purpose of using a *probabilistic* approach rather than the simple deterministic approach is to try to take into account the uncertainties….so that a more realistic analysis of the safety of a structure can be performed."

The probabilistic approach demands the accurate knowledge of the input probabilistic information. As Gertabakh (1980) writes:

"Computation of system reliability characteristics can be carried out only on the basis of data. Just as no one can cook a fine meal from poor ingredients, it is hopeless to try to derive accurate conclusions about system reliability without good data on the probabilistic properties of system elements, repair times, environmental conditions, etc."

One cannot divorce the search for the input probabilistic information from the probabilistic analysis. Elishakoff (1991) notes:

"It appears that the probabilistic analyst in actuality says: "Give me the probability densities of random variables or random functions involved and I will calculate the reliability of the structure!" This reminds us of the well-known statement by Archimedes: "Give me a firm spot on which to stand, and I will move the earth!"

**Safety fence around Mona Lisa (painted in 1503) in the Louvre Museum in Paris
(courtesy of Ing. Denis Meyer)**

**Leonardo da Vinci
(born in Florence on April 15, 1452, died in Cloux, France on May 2, 1519)**

So, where do we stand now, may ask an inquiring reader? Partial answer is provided by Rackwitz (1991), who writes:

"In 1981, A.C. Cornell, called structural reliability theory a healthy adolescent. We have seen that at that time, essential conceptual and methodological achievements have not yet seen a sight. Still, his evaluation of the state of development was fair as 90% of the practical problems already could have a solution. Now, 10 years more of developments have almost covered the remaining 10% if our demands were still those of 1981. But the increased knowledge necessarily has also increased the demands both from an academic and a practical aspect. It is certainly wise to be hesitant against generalizing statements that the academic reliability community is occupied more and more with problems which usually have little practical significance."

Combining the probabilistic methods with the traditional safety factor analysis may lead to the change in the safety factor allocations. What is the possible effect of such a change? Lind (1991) writes:

"Safety factors are the parameter settings of design codes. Neither the risk of failure nor the initial cost of the structures should be excessive. The safety factors should be selected such that the codes give the optimum service to society....A change in safety factors affects the productivity of the building sector and the safety of the product, and may thus affect GNP, life expectancy and other social indicators."

After the tragic events of September 11, 2001, set of issues was raised "regarding how we deal with events where these considerable ambiguity and uncertainty about the likelihood of their occurrence and their potential consequences" (Kunreuther, 2002). According to the latter study:

"A particularly startling feature of the September 11 attacks was the drama for disruption of the activities of the world's most powerful nation by a handful of determined individuals. This suggests that risk assessment needs to be supplemented by *vulnerability analysis* that characterizes the forms of physical, social, political, economic, cultural, and psychological harms to which individuals and modern societies are susceptible. Many millions of dollars have been spent on a variety of actions that are designed to reduce our vulnerability."

Papers by Slovic (2002), Deisler (2002), and Garrick (2002) address some of the issues associated with risk assessment to address terrorism. Theory of structural vulnerability, in conjunction with structural reliability, becomes of paramount importance. Mechanics community can do a lot for reducing the risk of terrorism. Reliability and vulnerability studies become centrally vital for the present and future of the world (Lind, 1995, 1996a, 1996b; Pinto, Blockley and Woodman, 2002).

Over a quarter century ago, Bolotin (1966) challenged the mechanics community:

"In conclusion, we would like to emphasize the importance of the theory of reliability in mechanics. The object of mechanics of a deformed solid is determination of the stresses and deformation in bodies. However, mechanics should also be concerned with how their findings are subsequently applied in structural analysis; with what–and in what form–is required for normative analysis and optimized design. It would be even better if mechanicists were to "invade" this field actively. Experience shows that in the past each such "invasion" proved fairly fruitful and made for a significant leap in the development and grounding of normative analysis. On the other hand, lagging behind in this field can--fortunately, only temporarily--deprive of practical value even the most brilliant achievement of mechanical scientists."

As if in the direct response to Bolotin's (1966) challenge, Oden, Belytschko, Babuška and Hughes (2003) write:

"During the next decade, probabilistic modeling of mechanical problems will be a topic of great importance and interest including stochastic features into computational models will not only provide realistic simulations of physical events but will also provide the analyst with specific information on the probabilities that can be assigned to predictions…. New methods for testing uncertainty will become important in virtually all branches of mechanics – fluid mechanics, mechanics of materials, solid mechanics; they will also promote the development of new computerized techniques to analyze uncertainty on engineering systems."

The present embracing of probabilistic methods is in a sharp contrast to the past as described by Pugsley: "When in autobiographical mood, I sometimes like to recall how, when I first showed signs of interest in possible statistical relations between safety and structural strength, my efforts not only received little support but were very nearly officially suppressed."

One should note that the concept of safety factor that emerged in applied mechanics found its way in medicine (Butterfield, 1963; Jones and Jones, 1982; Korol and Lang, 1965; Shiller and Stalberg, 1977; Taylor et al, 1982) and in business (Solomon et al, 1983) as a precaution factor. It is hoped that the methods discussed in this monogragh could be extended to these and other fields.

For those who did not have the patience to delve in the book but are looking for the answer to the question on the title of this book, it appears that the structural reliability and safety factors can peacefully coexist; they ought be treated as the friends rather than the foes; moreover, the concept of reliability may serve the framework that will allow for more rational allocation of the safety factors.

Bibliography

"Nothing is more instructive than a good bibliography."
G. Sarton (1952)

Abernethy R. B., Breneman J. E, Medlin C. H. and Reinman G. L., *Weibull Analysis Handbook,* Air Force Wright Aeronautical Laboratory Report, AFWAL-TR-83-2079, 1983.

Abramovich H., Singer J., and Yaffe R., Imperfection Characteristics of Stiffened Shells-Group1, *TAE Report 406*, Department of Aeronautical Engineering, Technion-Israel Institute of Technology, Haifa, Israel, 1981.

Administration, *Proceedings of the Reliability and Maintainability Conference,* IEEE Press, pp. 763-768 New York, 1966.

Aerospace Reliability and Maintainability Conference, pp. 278-283, AIAA Press, New York, 1963.

Agarwal J., Vulnerability Analysis of 3- Dimensional Structures, in *Structural Safety and Reliability* (Corotis R.B, ,Schuëller G.I. and Shinozaka M., eds.), p.4, Balkema, Lisse, 2001.

Agarwal J., Blockley D.I. and Woodman N.J., Vulnerability of Systems, Civil Engineering and Environmental Systems, Vol. 18, 141-165, 2001.

Akhmetov R.M. and Chertykovtsev V.K., Evaluation of the Professional Safety Factor, *Neftisnoe Khozaistvo*, No.9, 40-41, 1993.

Aitchison J. and Brown J.A.C., *The Lognormal Distribution*, Cambridge University Press, 1957.

Allan R., The Safety Factor in Test Equipment, *IEEE Spectrum,* 43-47, April, 1977.

Allen D.E., Safety Factors for Stress Reversal, *International Association for Bridge and Structural* Engineering *Publications,* Vol. 29(2), 19-27, 1969.

Allen D.E., Criteria for Design Safety Factors and Quality Assurance Expenditure, in Structural Safety and Reliability (Moan T. and Shinozuka M., eds.), *ICOSSAR '81*, Trondheim, Norway, pp. 667-678, 1981.

American Society of Metals, *Metals Handbook*, Vol. 1, Properties and Selection, 8th ed., 1966.

Ang A. H.-S. (Chairman, Task Committee on Structural Safety), Structural Safety – A Literature Review, *Journal of the Structural Engineering Division*, Vol. 98, 845-884, 1972.

Ang A. H-S. and Amin M., Reliability of Structures and Structural Systems, *Journal of Engineering Mechanics Division*, Vol. 94, 671-691, 1968.

Ang A. H-S. and Tang W. H., *Probability Concepts in Engineering Planning and Design*,Vol. 2, Wiley, New York, 1984.

Ang A. H-S. and Amin M., Safety Factors and Probability in Structural Design, *Journal of Structural Division*, Vol. 95, 1389-1404, 1969.

Ang A. H.-S. and Tang W. H., *Probability Concepts in Engineering, Planning and Design*, Vol. 1, Wiley, New York, 1979.

Ang A.H.-S. and De Leon D. Determination of Optimal Target Reliabilities for Design and Upgrading of Structures, *Structural Safety*, Vol. 19, 91-103, 1997.

Anonymous, Practical Reliability – Vol. I, "Parameter Variation Analysis", *NASA CR-1126*, 1968.

Anonymous, Practical Reliability – Vol. II, "Testing", *NASA CR-1128*, 1968.

Anonymous, Practical Reliability – Vol. IV, "Prediction", *NASA CR-1129*, 1968.

Anonymous, Symposium on Safety Factor in Vehicle Design, *British Medical Journal*, Vol. 2(5456), p.288, 1965.

Apostolakis G. and Kaplan S., Pitfalls in Risk Calculations, *Reliability Engineering*, Vol.2, 135-145, 1981.

Arbocz J., The Imperfection Data Bank: A Means to Obtain Realistic Buckling Loads, in *Buckling of Shells* (Ramm E., ed.), Spinger, Berlin, pp. 535-567, 1982.

Arbocz J., and Abramovich H., The Initial Imperfection Data Bank at the Delft University of Technology, Part 1, *Report LR-290*, Department of Aerospace Engineering, Delft University of Technology, Delft, The Netherlands, 1979.

Arbocz J., and Williams J. G., Imperfection Surveys on a 10 ft. Diameter Shell Structure, *AIAA Journal*, Vol. 15, 949-956, 1976.

Ardillon E., Barthelet B., and Sorensen J. D., Probabilistic Calibration of Safety Factors for Nuclear Operating Installations, *ASME Pressure Vessels, Piping Division Publication*, PVP Vol. 376, pp. 73-81, 1998.

Arone R.G., Statistical Evaluation of Strength of Metals in Brittle Fracture, *Problemy Prochnosti*, Vol. 3, No.1, 43-49, 1972 (In Russian).

Arora J. S., *Introduction to Optimum Design*, McGraw Hill, New York, 1989.

Ascher J., Comment on 'Constant, Failure Rate – a Paradigm in Transition' by J.A. McLinn, *Quality and Reliability Engineering International*, Vol. 7(5), 363-364, 1991.

ASME Pressure Vessels and piping Division, PVP, Vol. 386, pp. 25-32, ASME Press, New York, 1999.

Asplund S.O., The Risk of Failure, *Structural Engineer*, Vol.36,(8),1956.

Augusti G., Some Observations on Calculation of Structural Failure Probability, *Report UFIST/03/1974*, Department of Civil Engineering, University of Florence, 1974.

Augusti G., Baratta A. and Casciati F., *Probabilistic Methods in Structural Engineering*, Chapman and Hall, London, 1984.

Avakov V.A., Safety Factor in Fatigue under Fluctuating Stresses, *Journal of Vibration, Acoustics, and Reliability in Design*, Vol. 109, 397-401, 1987.

Awad A.M. and Gharrof M.K., Estimation of P(Y<X) in the Burr Case: A Comparative Study, *Commun. Statist.-Simulation Comput.*, Vol. 15, 389-403, 1986.

Ayyub B.M. and White G.J., Probability – Conditioned Partial Safety Factors, *Journal of Structural Engineering*, Vol. 113, 279-294, 1987

Bai Y. and Song R.X., Fracture Assessment of Dented Pipes with Cracks and Reliability-Based Calibration of Safety Factor, *International Journal of Pressure Vessels and Piping*, Vol. 74(3), 221-229, 1997.

Baker A.L.L., The Ultimate-Load Method of Design, *Concrete and Constructional Engineering*, Vol. 51(3), 293-302, 1956.

Baker M.J., The Evaluation of Safety Factors in Structures, *CIRIA Research Project 72, Final Report*, Dept. of Civil Engineering, Imperial College, London, UK., 1970.

Baker M.J., Variability in Strength of Structural Steel – A Study in Structural Safety, Part 1: Material Variability, *CIRIA Technical Note 44*, 1972.

Barlow R. E., Mathematical Theory of Reliability, *IEEE Transactions Reliability*, Vol. R-33, 16-20, 1984.

Barnett R.L. and Hermann P.C., Proof Testing in Design with Brittle Materials, *Journal of Spacecraft and Rockets*, Vol. 2(6), 956-961, 1965.

Basler E., Untersuchungen über den Sicherheitsbegriff von Bauwerken, *Schweizer Archiv*, 133, 1961 (in German).

Bedford T. and Cooke R., *Probabilistic Risk Analysis*, Cambridge University Press, 2001.

Beeby A.W., Partial Safety Factors for Reinforcement, *The Structural Engineer*, Vol. 72(20), 341-343, 1994.

Belenia E.I., Foreword, on *Selected Works* by N.S. Streletskii (Belenia E.I., ed.), pp. 5-7, "Stroiizdat" Publishers, Moscow, 1975 (in Russian).

Beliaev B.I., About Improving Principles of Determining Structural Reliability, *Stroitelnaya Mekhanika i Raschet Sooruzhenii*, No.5, 1974 (in Russian).

Bélidor B. F., de, *La science des ingénieurs dans la conduite des travaux de fortification et d'architecture civile*, Paris, 1729 (in French).

Ben-Haim Y., A Non-Probabilistic Concept of Uncertainty, *Structural Safety*, Vol. 14, 227-245, 1994.

Ben-Haim Y., A Non-Probabilistic Measure of Reliability of Linear Systems Based on Expansion of Convex Models, *Structural Safety*, Vol. 17, 91-109, 1995a.

Ben-Haim Y., Author's Reply on the Discussion on "A Non-Probabilistic Concept of Reliability", *Structural Safety*, Vol. 17, 198-199, 1995b.

Ben-Haim Y., *Robust Reliability in the Mechanical Sciences*, Springer, Berlin, 1996.

Ben Haim, Y., Must Reliability be Probabilistic?, *Journal of Statistics and Mathematical Simulation*, Vol.55(3), 263-265, 1996.

Ben-Haim, Y., Robust Reliability of Structures, in *Advances in Applied Mechanics* (Hutchinson J.W. and Wu T.Y., eds.), Vol.33, Academic Press, San Diego, pp.1-41, 1997.

Ben-Haim Y., *Information-Gap Decision Theory*, Academic Press, San Diego, 2001.

Ben-Haim Y., Design for Reliability: The Information-Gap Approach, in *Structural Safety and Reliability* (Corotis R.B., Schuëller G.I.and Shinozuka M., eds.), p.31, Balkema, Lisse, 2001.

Ben-Haim Y. and Elishakoff I., Convex Models of Vehicle Responde to Uncertain but Bounded Terrain, *Proceedings of the ASME Pressure Vessels and Piping Conference* (Chung H., ed.), Honolulu, Hawaii, ASME, 81-88, 1989a.

Ben-Haim Y. and Elishakoff I., Dynamics and Failure of a Thin Bar With Unknown but Bounded Imperfections, in *"Recent Advances Impact Dynamics of Engineering Structures"* (Hui D. and Jones N., ed.), AMD- Vol. 105, ASME, New york, 89-96, 1989b.

Ben-Haim Y. and Elishakoff I., Non-Probabilistic Models of Uncertainty in the Non-linear Buckling of Shells with General Imperfections: Theoretical Estimates of the Knockdown Factor, *Journal of Applied Mechanics*, Vol. 111, 403-410, 1989c.

Ben-Haim Y. and Elishakoff I., *Convex Models of Uncertainty in Applied Mechanics*, Elsevier Science Publishers, Amsterdam, pp. 1-43, 1990.

Benjamin, J. R. and Cornell, C. A., *Probability, Statistics and Decision for Civil Engineers*, McGraw Hill, New York, 1970.

Benjamin J.R., Schuëller G.I. and Witt F.J., A Critical Review of the Results of the Second International Seminar on Structural Reliability of Mechanical Components and Subassemblies of Nuclear Power Plants, *Reliability Engineering*, Vol. 2, 125-134, 1981.

Berg A.I., Science about Reliability, *Economical Gazette*, June 8, 1961 (in Russian).

Berg H. P., Grundler D. and Lange F., Risk Comparison: One Way to Risk Evaluation, *6th ICOSSAR*, Vol. 3, pp. 1847-1943, 1993.

Bernshtein S.A., *Essays on History of Structural Mechanics*, State Publishing House on Civil Engineering and Architecture, Moscow, 1957 (in Russian).

Bhattacharyya G.K. and Johnson R.A., Stress-Strength Models for Reliability: Overview and Recent Advances, *Proceedings 26th Design of Experiments Conference*, pp.531-548, 1981.

Bieber T., Private Communication, July 25, 2003.

Bier V.M., Challenges of the Acceptance of Probabilistic Risk Analysis, *Risk Analysis*, Col.19, 703-710, 1999.

Bilikam J.E., Some Stochastic Stress- Strength Processes, *IEEE Transactions on Reliability*, Vol R-34, No.3., 269-274, 1985.

Bienaymé I- J., Considérstions à l'appui de la découverte de Laplace sur la loi probabilité dans la méthode des moindres carrés *Comptes Rendy*

Birger I. A., Probability of Failure, Safety Factors and Diagnostics, *in Problems of Mechanics of Solid Bodies*, "Sudostroenie" Publishers, Leningrad, pp. 71-82, 1970 (in Russian).

Birger I. A., Deterministic and Statistical Models of Fatigue Strength, *Strength of Materials*, Vol. 14 (4), pp. 444 - 450, 1982.

Birnbaum Z.W., On the Use of Mann-Whitney Statistics, *Proceedings, Third Berkeley Symposium in Mathematical Statistics and Probability*, Vol. 1, 13-17,Unversity of California Press, 1956.

Birnbaum Z.W. and McCarty R.C., A Distribution-Free Upper Confidence Bound for $P(X<Y)$ Based on Independent Samples of X and Y, *Ann.. Math. Satatist.*, Vol. 29, 558-562, 1958.

Blockley D.I., *The Nature of Structural Design and Safety*, Ellis Horwood Ltd., Chichester, 1980.

Blockley D.I., Engineering from Reflective Practice, Research in Engineering Design, Vl.4, 13-22, 1992.

Blockley D.I. and Godfrey P., *Doing It Differently*, Thomas Telford, 2000.

Bodner S., Private Communication, September, 2003.

Bolotin V.V., Mechanics of Solid Bodies and the Theory of Reliability, in *Mechanics of Solids*, Proceedings of the Second All-Union Congress on Theoretical and Applied Mechanics, Vol.3, 68-82, 1966 (in Russian).

Bolotin V.V., Modern State of the Art of Reliability and Statistical Mechanics of Structures, in *Problems of Reliability in Structural Mechanics*, Vilnius, pp. 7-15, 1968 (in Russian).

Bolotin V.V., *Statitical Methods in Structural Mechanics*, Holden Day, San Francisco, 1969.

Bolotin V. V., *Application of the Methods of the Theory of Probability and the Theory of Reliability to Analysis of Structures*, State Publishing House for Buildings, Moscow, 1971, (in Russian). English translation: FTD-MT-24-771-73, Foreign Technology Div., Wright Patterson, AFB, Ohio, 1974.

Bolotin V. V., *Wahrscheinlichkeitsmethoden zur Berechnung von Konstruktionen*, VEB Verlag von Wilhelm Ernst & Sohn, Berlin, 1981 (in German).

Bolotin V.V., Goldenblat an Smirnov A.F., *Modern Problems of Structural Mechanics*, Izdatelstvo Literatury po Stroitelstvu, Moscow, 1964 (in Russian).

Bolotin V.V., Goldenblat I.I. and Smirnov A.F., *Structural Mechanics: Modern State of the Art and Development Perspectives*, Second edition, Izdatelatvo Literatury po Stroitelatvu, Moscow, 1972 (in Russian).

Bompas-Smith J.H., The Determination of Distribution that Describe the Failure of Mechanical Components, *Eighth Annals of Reliability and Maintainability*, pp. 343-356, 1969.

Bompas Smith J.H., in *Mechanical Survival: The Use of a Reliability Data* (Brook R.H.W., ed.), Mc Graw-Hill, New York, 1973.

Borch K.H., *The Economics of Uncertainty*, Princeton University Press, 1968.

Borges J.F. and Castanetha M., Structural Safety, Laborstorio Naciohal de Engenharia Cival, Lisboa, Portugal, 1971.

Breitung K., Asympotitic Approximations for Multinormal Integrals, *Journal of Engineering Mechanics*, Vol.110, 357-366, 1984.

Breitung K., Parameter Sensitivity of Failure Probabilities, *in Proc. 3rd IFIP WG 7.5 Working Conference* (Der Kiureghian A., and Thoft-Christensen P.,eds), Berkeley, California, pp.43-51, Springer, Berlin,1990.

Breugel K. Van, Acceptance Criteria for High Consequence Risk: A Critical Appraisal, *6th ICOSSAR*, Innsbruck, pp. 1849-1856, 1993.

Breugel K.Van, How to Deal and Judge the Numerical Results of Risk Analysis, Computers and Structures, Vol. 67, 157-164, 1998.

Brinch-Hansen J., Limit Design and Partial Safety Factor in Soil Mechanics, *Danish Geotechnical Institute Bull.*, Vol.1, P.4, 1956.

Broam C.B., Concepts of Structural Safety, *Journal of Structural Division*, Vol. 83, 39-57, 1960.

Broding W.C. Diedrich F.W. and Parker P.S., Structural Optimization and Design Based on Reliability Design Criterion, *Journal of Spacecraft and Rockets*, Vol. 1, 56-61, 1964.

Brown C.B.,Concepts of Structural Safety, *Journal of the Structural Division*, Vol.86(12),1960.

Brown K.A.P., In Pursuit of Reliability, *Reliability Engineering*, Vol. 10, 141-150, 1985.

Bruevich N.G. and Grabovetsky V.P. On the Basic Directions Taken in Reliability Theory, *Kibernetica na Sluzhbu Kommunizmu,"* Energya" Publishing, Moscow, 1964 (in Russian).

Bruhn E.F.(Editor-in-Chief), *Analysis and Design of Missile Structures*, Tri-State offset Company, Cincinnati, OH, 1975.

Bucher C.G, Adaptive Sampling – an Iterative Fast Monte Carlo Procedure, *Structural Safety*, Vol. 5(3), 119-126, 1988.

Bucher C.C and Schuëller G.I., Software for Reliability Based Analysis, *Journal of Structural Safety*, Vol.16, 13-22, 1994.

Bulychev A.P., Reliability of Structures with Finite Number of Random Parameters due to Time-Varying Random-Excitation, in *Loads and Reliability of Building Constructions*, Proceedings, ZNIISZK, Vol. 21, Moskow 1973a (in Russian).

Bulychev A.P., Some Reliability Questions of Building Constructions, in *Load and Reliability of Building Constructions,* Proceedings, ZNIISK, Vol.21, Moscow, 1973b (in Russian).

Bulychev A.P. and Sukhov Yu.D., Use of the Theory of Reliability for Code Making of Design Values of Loads, *Stroitelnaya Mekhanika i Raschet Sooruzhenii*, 15-19, 1973 (In Russian).

Burros R.H., Probability of Failure of Building from Fire, *Journal of Structural Division*, Vol. 101, 1947-1960, 1975.

Bushnell D., GENOPT-A Program That Writes User Friendly Optimization Code, *International Journal of Solids and Structures*, Vol. 26 (9-10), 1173-1210, 1990.

Bushnell D., Private Communication, August 15, 2003.

Bussiere R., Predict Reliability First... Build It Later, *SAE Journal*, Vol. 69 74-76, 1961.

Butterfield B., From Donor to Patient- Another Safety Factor, Transfusion, Vol.3(2), p. 125, 1963.

Cable C.W. and Virene F.P., Structural Reliability with Normally Distributed Static and Dynamic Loads and Strength, *Proceedings of the Annual Symposium on Reliability*, pp. 329-336, IEEE Press, New York, 1967.

Camp B.H., Generalization to N-Dimensions of the Inequalities of Chebychev Type, *Annals of Mathematical Statistics*, Vol. 19, 568-574, 1948.

Campbell C.C., Safety Margins Established by Combined Environmental Test Increase Atlas Missile Component Reliability, *IRE Transactions on Reliability and Quality Control*, Vol.10, 1-6, 1961.

Carpenter R.B. Jr., Apollo Reliability by Demonstration or Assessment, *Proceedings of Tenth National Symposium on Reliability and Quality Control*, pp.517-524, IEEE Press, New York, 1964.

Carter A.D.S., Reliability Reviewed, *Proc.Inst. Mech.Eng.,*Vol.193,1979.

Carter A.D.S., *Mechanical Reliability*, Macmillan, Basingstoke, 2nd ed., 1986.

Carter A.D.S., *Mechanical Reliability and Design*, Wiley, 1997.

Casciati F., Safety Index, Stochastic Finite Elements and Expert Systems, in *Reliability Problems* (Casciati F. and Roberts, J.B eds.), pp 51-88, Springer, Vienna, 1991.

Casciati F. and Faravelli L., Load Combination by Partial Safety Factors, *Nuclear Engineering and Design*, Vol. 75, 439-452, 1982.

Casciati F., and Roberts J. B., eds., Reliability Problems: *General Principles and Applications in Mechanics of Solids and Structures*, Springer, Vienna, 1991.

Castellani A., Average Value of the Safety Coefficient in a Framed Structure Corresponding to Random Distribution of Yield Stresses, *Meccanica*, 281-285, Dec 1969.

Castellani A., Safety Margins of Suspension Bridges Under Seismic Conditions, *Journal of Structural Engineering*, Vol.133, 1600-1616, 1987.

Cathey B.H., Pressure System Safety Factors Made Simple, *Professional Safety*, 32-35, Feb.1978.

Cawley J.C., A Statistical Determination of Spark Ignition Safety Factor in Methane, Propane and Ethylene Mixture in Air, *Report No. 9048*, Interior Bureau of Mines, United States Department of Interior, Washington D.C., 1986.

Cederbaum G., Elishakoff I., and Librescu L., Reliability of Laminated Plates via the First-Order Second-Moment Method, *Journal of Composite Structures*, Vol.15, 161-167,1990.

Cempel C., *Vibroacoustical Condition Monitoring*, Ellis Horwood, Chichester, England, 1991.

Cesare M.A. and Sues R.H., ProFES, Probabilistic Finite Element Software–Bring Probabilistic Mechanics to the Desktop, AAIA Paper 99-1607, 1999.

Chebychev P.L. Sur les valeurs limites des integrals, *Liouville's J. Math. Pures et Appl.*, Vol. 19, 157-160, 1874 (in French).

Cheen W.K., Some Stress-Strength Reliability Models, *Microelectronic Reliability*, Vol. 22(2), 227-283, 1982.

Chen G. H., and Dai S. H., Determination of Partial Safety Factors of Parameters for Integrity Assessments of Welded Structures Contraining Defects, *International Journal of Pressure Vessels and Piping*, Vol. 72, 19-25, 1997.

Chen Y.M. and Bougund U., Statistical Analysis of Observed Materials and Structural Data – A Use's Manual, *Institute of Engineering Mechanics, University of Innsbruck*, 1986.

Chenevert M.E., Microcomputer Program Helps Determine Kick Safety Factor, *World Oil*, 62-66, Dec. 1983.

Chernenko V.I., Reliability Safety Factor of Tube-Tube Wall Joints in Heat Exchangers, *Soviet Energy Technology*, No.5, 35-39, 1984.

Chilver A.H.,Some Problems of Structural Safety, *British Welding Journal*, Vol.2(8),1955.

Chou K., McIntosh C. and Corotis R.B., Observations on Structural System Reliability and the Role of Modal Correlations, *Structural Safety*, Vol. 1, 189-198, 1983.

Church J.D. and Harris B., The Estimation of Reliability from Stress-Strength Relationships, *Technometrics*, Vol. 12, 49-54, 1970.

Churchley A. R., A Rationale for the Reliability Assessment of High Integrity Mechanical Systems, *Reliability Engineering*, Vol.19, 59-71, 1987.

Clausen R.J., Safety Factor in Car Design, British Medical Journal, Vol. 1 (May 4), p.1065, 1957.

Close E.R., Beard L.R. and Dawdy D.R., Objective Determination of Safety Factor in Reservoir Design, *Journal of Hydraulic Division*, Vol. 96, 1167-1177, 1970.

Cohen H., Space Reliability Technology: A Historical Perspective, *IEEE Transactions on Reliability*, Vol. R-33, No. 1, 36-40, 1984.

Colquhon I. Menendez A. and Dovico R., Method Yields Safety Factor for In-line Inspection Data, Oil and Gas Journal, Vol. 96(38), 83-86, 1998.

Cook R. and Bedford T., Reliability Databases in Perspective, *IEEE Transactions on Reliability*, Vol.51, 294-310, 2002.

212

Cooke R.M., The Design of Reliability. Data Bases – Parts i and ii, *Reliability Engineering and System Safety*, Vol.51, 137-146, 209-223, 1996.

Cooke R., Bedford T., Meilijson I. and Meester L., Design of Reliabilit Data Bases for Aerospace Applications, *European Space Agency*, Department of Mathematics, Delft University of Technology, Technical Report 93-110, 1993.

Cornell, C. A., Probability-based Structural Code, *ACI Journal*, Vol. 66, 974-985, 1969.

Cornell A., Probabilistic Seismic Hazard Analysis: A 1980 Assessment, *7WCEE*, Istanbul, 1980a.

Cornell C.A., Some Thoughts on Systems and Structural Reliability, *Nuclear Engineering and Design*, Vol. 60, 115-116, 1980b.

Cornell C.A., Utilization of Present Knowledge of Probabilistic Structural Reliability in Analysis of Nuclear Power Plants, *Nuclear Engineering and Design*, Vol. 60, 33-36, 1980c.

Cornell, C. A., Structural Safety: Some Historical Evidence That It is a Healthy Adolescent, *Structural Safety and Reliability* (Moan T. and Shinozuka M., eds.), Elsevier Scientific Publishing Company, Amsterdam, 19-29, 1981.

Cramér H., *Mathematical Method of Statistics*, Princeton University, New Jersey, pp. 131-134, 221-231, 1946.

Cramér H., *The Elements of Probability Theory and Some of its Applications*, Wiley, New York, pp. 116-118, 1965.

Cramér H., *Random Variables and Probability Distributions*, Cambridge University Press, p.87, 1970.

Cross N., *Engineering Design Methods*, Wiley, 2nd ed., Chichester, 1994.

Crowley V.F., The Use of Safety factor in Transmission Line Design, *Transactions of the Canadian Electrical Association, Engineering and Operating Division*, Vol 15, Part 3, 1-10, 1976.

Cruse T. A. (ed.), *Reliability-Based Mechanical Design*, Marcel Dekker, New York, 1997.

Curback M., Jesse F. and Proske D., Partial Safety Factor for Textile Reinforced Concrete, in *Structural Safety and Reliability* (Corotis R.B.,Schuëller G.I.and Shinozuka M., eds.),p.63, Balkema, Lisse, 2001.

Curnick G.E., The Effect of Proof-Load on the Reliability of a Structural System, M.A. Sc. Thesis, University of Waterloo, Canada, 1970.

Dandrea R.A. and Sangrey D.A., Safety Factors for probabilistic Slope Design, *Journal of Geotechnical Engineering Division*, Vol 108, 1101-1118, 1982.

Danieli M., Private Communication, July 24, 2003.

Danilevsky A., Safety Factor Dams and Retaining Walls, *Journal of the Geotechnical Engineering Division*, Vol 108, 47-61, 1982.

Dao-Thien M. and Mossoud M., On the Relation between the Factor of Safety and Reliability, *Journal of Engineering for Industry*, Paper No. 73-WA/DEI.

Dargahi-Noubary G.R., On Estimation of failure Probability, *Reliability Engineering and System Safety*, Vol. 23, 23-30, 1988.

Dargahi-Noubary G.R., On Tail Estimation- An Improved Method, *Mathematical Geology,* Vol.21, 829-842, 1990.

Dargahi-Noubary G.R., On Reliability Calculations when Stresses are Generated by a Non-Homogeneous Poisson Process, *Reliability Engineering and System Safety*, Vol. 31, 255-263, 1991.

Davies G.R., Safety Factor in Car Design, British Medical Journal, Vol. 1 (April 20), p.949, 1957.

De Beer E., Lousberg E., DeJonghe A., Wallayo M. and Carpentier R., Partial Safety Factors in Pipe Bearing Capacity, *Proceedings of the 10 th International Conference on Soil Mechanics and Foundation Engineering*, 105-110, June 1981.

Deisler P.F., Jr., A Perspective: Risk Analysis as a Tool for Reducing the Risks of Terrorism, *Risk Analysis*, Vol.22, 405-413, 2002.

De Mollerat T. and Vidal C., Evaluation Design and Tests Safety Factors, *Final Report of ESTEC Contract No. 6370/85/NL/PB*, Cannes,1986.

Demonsablon P. and Jouanna P., Research Determining the Safety Factor of Dam Slopes by Digital Computer, *Proceedings of the 10th International Congress on Large Dams*, Montreal, Canada, pp.11–26, June 1970.

Der Kiureghian A., Shang Y. and Li C.C., Inverse Reliability Problem, *Journal of Engineering Mechanics*, Vol.120, 1154-1159, 1994.

Det Norke Veritas, Safety, *Man and Society*, D.N.V., 1322, Hølvik, Norway, 1977.

Dhillon B.S., *Human Reliability: With Human Factors*, Pergsman Press, New York, 1968.

Dhillon B.S., Mechanical Reliability: Interference Theory Models, *Proceedings of the Annual Reliability and Maintainability Symposium*, pp.462-467, 1980a.

Dhillon B.S., Stress/Strength Reliability Models, *Microelectronic Reliability*, Vol. 20, 513-516, 1980b.

Dhillon B.S., Bibliography of Literature on Transit System Reliability, *Microelectronics and Reliability*, Vol. 22, 641-651, 1982.

Dhillon B.S., Stochastic Models for Evaluating Probability of System Failure due to Human Error, *Microelectronics and Reliability*, Vol. 24, 921-924, 1984.

Dhillon B.S., *Human Reliability: With Human Factors*, Pergamon Press, New York, 1986.

Dhillon B.S., *Mechanical Reliability: Theory, Models and Applications*, AIAA Press, Washington D.C, 1988.

Dhillon B. S., Bibliography of Literature on Safety Factors, *Microelectron. Reliab.*, Vol. 29 (2), 267-280, 1989.

Dhillon B.S. and Singh C., *Engineering Reliability: New Techniques and Applications*, Wiley, New York, 1981.

Dhillon B.S., Proctor C.L. and Elsdyed E.A.R., Stress-Strength Reliability Analysis of Redundant System, *Reliability, Stress Analysis and Failure Prevention Methods in Mechanical Design* (Milestone W.E., ed.), pp. 9-12, ASME Press, New York, 1980.

Disney R.L., Lipson C. and Sheth N.J., The Determination of the Probability of Failure by Stress/Strength Interference Theory, *Proceedings of 1968 Annual Symposium on Reliability*, IEEE, New York, 1968.

Ditlevsen O., *Uncertainty Modeling with Applications to Multidimensional Civil Engineering Systems*, Mc. Graw-Hill, New York, 1981.

Ditlevsen O., The Fake of Reliability Measures as Absolutes, *Nuclear Engineering and Design*, Vol. 71, 439-440, 1982.

Ditlevsen O., Fundamental Postulate in Structural Reliability, *Journal of Engineering Mechanics*, Vol. 109, 1096-1102, 1983.

Ditlevsen O., Probabilistic Thinking – An Imperative in Engineering Modelling: An Essay, *Report No. 192*, Series R, Department of Structural Engineering, Technical University of Denmark, 1984.

Ditlevsen O., *Structural Reliability Methods*, SBI, 1990 (in Danish).

Ditlevsen O., Uncertanty and Structural Reliability: Hocus Pocus or Objective Modelling, *Report No. 226, Serie R*, Department of Structural Engineering, Technical University of Denmark, 1988.

Ditlevsen O. and Madsen H.O., *Structural Reliability Methods*, Wiley, Chichester, 1996.

Dolinski K., First-Order Second-Moment Approximation in Reliability of Structural Systems: Critical Review and Alternative Approach, *Structural Safety*, Vol. 1, 211-231, 1983.

Donovan N. C. and Bornstein A. E., Uncertainties in Seismic Risk Procedures, *Journal Geotechnical Engineering Division*, Vol. 104, 869-887, 1978.

Doroshchuk G.P., Towards Probabilistic Justification of Coded Stress Analysis, *Stroitelnaya Mekhanika i Raschet Sooruzhenii*, No. 4,1974 (in Russian).

Downson F., The Estimation of $Pr(Y<X)$ in the Normal Case, *Technometrics*, Vol. 15, 551-558, 1973.

Dresden M., Chaos: A New Scientific Paradigm or Science by Public Relations?, *The Physics Teacher*, Vol. 30, 10-14 and 74-80, 1992.

Driving A.Ya., Ditlevsen O., Wiley, Probabilistic-Economic Method in the Code Marking of Building Constructions, *Stroitelnaya Mekhanika i Raschet Sooruzhenii*, No. 3, 7-11, 1982 (in Russian).

Drucker D.C., Greenberg H.J. and Prager W., The Safety Factor of an Elastic-Plastic Body in Plane Strain, *Journal of Applied Mechanics*, Vol. 18(4), 371-378,1951.

Dulacska E., The Safety Factor to be Applied in Shell Buckling Analysis, *Acta Tachnica Acad., Sci. Hung.*, Vol. 99, 9-30, 1986.

Dummer G.W.A., The Reliability of Components in Satellites, *Journal of the British Institution of Radio Engineers*, Vol. 21, 457-463, 1961.

Dyrobe C., Gravesen S., Krenk S.,Lind N. and madsen H.O., *Konsfruktioners Sikkerhed*, Danmarks Tekniske Hjskole, Kbenhavn, 1983 (in Danish).

Ebel G. and Lang A. Reliability Approach to the Spare Parts Problem, *Proceedings of Ninth Symposium on Reliability and Quality Control*, pp. 85-92, IEEE Press, New York, 1963.

Ebrahimi N., On Estimating the Reliability for a Stress-Strength System, *IEEE Transactions on Reliability*, Vol. R-34, No. 3, p. 280, 1985.

Editorial, Alexander Moiseevich Kakushadze, *Stroitelnaya Mekhanika i Raschet Sooruzhenii*, No. 3, p. 61, 1974 (in Russian).

Edgeworth F.Y., *Mathematical Psychics*, Kegan Paul, Long, 1881.

Edgeworth F.Y., *Papers Relating to Political Economy*, 3 volumes, Macmillan, 1925.

Editorial, Gábor Kazinczy (1889-1964), *Acta Technica Academiae Scientiarum Hungaricae*, Vol. 53, 455-460, 1966, (in German).

Edlund B. and Leopoldsen U., *Scatter in Strength of Data of Structural Steel*, Publication 72:4, Department of Structural Engineering, Chalmers University of Technology, 1980.

Eibinder S.K., Reliability Models and Estimating in Terms of Stress-Strength System, *Ph.D. Dissertation*, Polytechnic Institute of New York, New York, 1968.

Ekimov, V. V., *Probabilistic Methods in the Structural Mechanics of Ships*, "Sudostroenie" Publishing House, Leningrad, 1966 (in Russian).

Elishakoff I., How to Introduce Initial-Imperfection Sensitivity Concept into Design, in *Collapse: The Buckling of Structures* (Thompson J. M. T., and Hunt G. W., eds.), Cambridge University Press, pp. 345-357, 1983.

Elishakoff I., *Probabilistic Theory of Structures*, Dover, New York, 1999 (first edition: Wiley, 1983).

Elishakoff I., Simulation of an Initial Imperfection Data Bank, Part 1: Isotropic Shells with General Imperfections, TAE Report 500, Department of Aeronautical Engineering, Technion-Israel Institute of Technology, Haifa, Israel 1982. Also in *Buckling of Structures-Theory and Experiment* (Elishakoff I., Arbocz J., Babcock C. D. Jr., and Libai A., eds.), Elsevier Science Publishers, Amsterdam, pp. 195-210, 1988.

Elishakoff I., Essay on Reliability Index, Probabilistic Interpretation of Safety Factor and Convex Models of Uncertainty, in *Reliability Problems*, (Cascisti F. and Roberts, J.B eds.), pp 237-271, Springer, Vienna, 1991a.

Elishakoff I., Probabilistic Models of Structural Components: A General Report, in *Sixth International Conference on Applications of Statistics and Probability in Civil Engineering*, (Esteva L., and Ruiz S.E., eds.), Vol.3,pp.126-132, Mexico City, 1991b.

Elishakoff I., Discussion on "A Non-Probabilistic Concept of Reliability", *Structural Safety*, Vol. 17, 195-198, 1995.

Elishakoff I.,Personal Communication to Prof. Y. Ben-Haim, Jan.9,1996.

Elishakoff I., How to Introduce Initial-Imperfection Sensitivity Concept into Design 2, pp. 237-267, *NASA/CP-1998-206280*, 1998.

Elishakoff I., Are Probabilistic and Antioptimization Methods Interrelated?, in *Whys and Hows in Uncertainty Modelling* (Elishakoff I., ed.), pp.285-317, Springer, Vienna, 1999a.

Elishakoff I., What May Go Wrong with Probabilistic Methods?, in *Whys and Hows in Uncertainty Modelling* (Elishakoff I., ed.), pp. 265-284, Springer, Vienna, 1999c.

Elishakoff I., Application of the Bienaymé and Tchebycheff Inequalities for the 'Structural Reliability' and 'Engineering Planning and Design' Courses, *International Journal of Mechanical Engineering Education*, Vol. 28(3),187-194,2000.

Elishakoff I., Interrelation between Safety Factor and Reliability, *NASA CR-2001-211309*, 2001.

Elishakoff I. and Ben-Haim, Y., Dynamics of a Thin Cylindrical Shell under Impact with Limited Deterministic Information on Its Initial Imperfections, *Journal of Structural Safety*, 1990.

Elishakoff I. and Cederbaum G., Contrasting Some Exact and Approximate Solutions for Reliability of Structure, in *Sixth International Conference on Applications of Statistics and Probability in Civil Engineering*, (Esteva L., and Ruiz S.E., eds.), Mexico City, pp.250-256, 1991.

Elishakoff I. and Colombi P., Combination of Probabilistic and Convex Models of Uncertainty when Scarce Knowledge is Present on Acoustic Excitation Parameters, *Computer Methods in Applied Mechanics and Engineering*, Vol. 104, 187-209, 1993.

Elishakoff I. and Hasofer A. M., On the Accuracy of Hasofer-Lind Reliability Index, *Proceedings of ICOSSAR-85, International Conference on Structural Safety and Reliability*, Kobe, Japan, Vol. 1, 229-239, 1985.

Elishakoff I. and Hasofer A. M., Detrimental or Serendipitous Effect of Human Error on Reliability of Structures, *Computer Methods in Applied Mechanics and Engineering*, Vol. 129, 1–7, 1996.

Elishakoff I. and Hasofer A. M., Exact versus Approximate Analysis of Structural Reliability, *International Journal of Thermal*, Mechanical, and Electromagnetic Phenomena in Continua, Vol. 44, 303 –312, 1987.

Elishakoff I. and Li Q, How to Combine Probabilistic and Antioptimization Methods?, in *Whys and Hows in Uncertainty Modelling* (Elishakoff I., ed.), pp. 319-339, Springer, Vienna, 1999b.

Elishakoff I., Li Y.W., and Starnes J.H., Jr., *Non-Classical Problems in the Theory of Elastic Stability*, Cambridge University Press, New York, 2001.

Elishakoff I. and Nordstrand T., Probabilistic Analysis of Uncertain Eccentricities on a Model Structure, in *Sixth International Conference on Applications of Statistics and Probability in Civil Engineering*, Mexico, 184-192, 1991.

Elishakoff I. and Starnes J.H.Jr, Safety Factor and the Non-Deterministic Approaches. *Proceedings of the 40th AAIA/ASME/ASCE/AHS/ASC Structures*, Structural Dynamics and Materials Conference, St. Louis, MO, 1999.

Elishakoff I., Gana-Shvili Y., and Givoli D., Convex Optimization as Applied to Uncertain Eccentricities, in *Sixth International Conference on Applications of Statistics and Probability in Civil Engineering*, Mexico, pp. 150-157, 1991.

Ellingwood B., Ley-Endecker E.V. and Yao J.T.P., Probability of Failure from Abnormal Load, *Journal of Structural Engineering*, Vol. 109, 875-890, 1983.

Ellingwood B., Galambos Th.V., Macgregor J.G. and Cornell C.A., Development of a Probability Based Load Criterion for American National Standard A58, *Special Publications 577*, National Bureau of Standard, June 1980.

Ellingwood B. and Galambos Th. V., General Specifications for Structural Design Loads, *Probabilistic Methods in Structural Engineering* (M. Shinozuka and J.T.P. Yao, eds.), ASCE, pp. 27-42, 1981.

Ellingwood B. and Shaver J.R., Reliability of RC Beams Subjected to Fire, *Journal of Structural Division*, Vol, 103, 1047-1059, 1977.

Engineering Process, *Proceedings of the 6th Advances in Reliability* Technology Symposium, pp. 279-296, National Center of Systems Reliability, Warrington, UK, 1980.

Enis P. and Geisser S., Estimation of the Probability that X<Y, *Journal of the American Statistical Association*, Vol. 66, 162-168, 1971.

Encyclopedia Brittanica, Supplement to the 4th, 5th, and 6th editions, article 'Coulomb', Vol. 3, pp. 414-419, 1824.

Esary J.D., Proschan F. and Walkup D. W., Association of Random Variables with Applications, *Ann. Mathematical Statistics*, Vol. 38, 1466-1474, 1967.

Esteva L. and Rosenblueth E., Use of Reliability Theory in Building Codes, in *Applications of Statistics and Probability to Soil and Structural Engineering*, Hong Kong, 1971.

Estler W.T., The Bienayme-Chebyshev Inequality, *Journal of Research of the National Institue of Standards and Technology*, Vol.102(5), 588, 1997.

Evans A.G. and Widerhorn S.M., Proof Testing of Ceramic Materials: An Analytical Basis for Failure Prediction, International Journal of Fracture, Vol.10(3), 379-392, 1974.

Evans M., Hastings N. and Pieacock B., *Statistical Distributions*, Wiley, Chichester, 1993.

Evans T.M. and Merkord D. Is There Life After 10,000 Flight Hours?, *Proceedings of Annual Reliability and Maintainability Symposium*, pp. 396-401, IEEE Press, New York, 1985.

218

Evsadze S., Private Communication, July 24, 2003.

Evsadze S., Private Communication, August 25, 2003.

Fadale T. and Sues R., Reliability-Based Analysis and Optimal Design of an Integral Airframe Structure lap Joint, AIAA Paper, *40th AIAA/ASCE/AHS/ASC Structures, Structural Dynamics and Materials Conference*, St. Louis, Mo, 12-15, April, 1999.

Faulkner D., Safety Factors?, *Steel Plated Structures, An International Symposium* (P. J. Dowling, J. Z. Harding and P. A. Fieze, eds.), Crosby Lockwood Staples, London.

Feld J., *Construction failures*, Wiley, New York, 1966.

Feng Y.S., Enumerating Significant Failure Modes of a Structural System by Using Criterion Method, *Computers and Structures*, Vol.30(5),1153-1157,1988a.

Feng Y.S., Structural System Reliability Combining the Constraint of damage tolerance Design, *Computers and Structures*, Vol.30(6), 1341-1346,1988b.

Feng Y.S., The Computation of Failure Probability for Nonlinear Safety Margin Equations, *Reliability Engineering and System Safety*, Vol.27, 323-331,1990.

Feng Y.S. and Song B.F., Reliability Analysis and Design for Multi-Box structures, *Computer and Structures*, Vol. 37(4), 413-422, 1990.

Ferry Borges J., Implementation of Probabilistic Safety Concepts in International Codes, in *Proceedings, Second International Conference on Structural Safety and Reliability*, (Kupfer H., Shinozuka M. and Schuëller, G.I., eds.), pp. 121-133, Werner Verlag, Düsseldorf, 1977.

Ferry Borges J. and Castanheta M., *Structural Safety*, 2nd ed., National Civil Eng. Lab., Lisbon, Portugal, 1971.

Feynman R. P., *What Do You Care What Other People Think?*, Bantam Books, New York, 1985.

Fischer D.H., *Historian's Fallacies*, Harper & Row, New York, pp. 3-8, 1970.

Fischer D. S., and Kavanaugh S., The Impact of Safety Factors on Pump Performance and Selection: Case Study, a Methodology for Correcting Pump Oversizing, *ASHRAE Transitions*, Vol.107, Part 2, 584-589, 2001.

Flint E., The Derivation Of Safety Factor For Design Highway Bridges, *The Design of Steel Bridges* (Rockey K.C and Evans H.R., eds.), Granada, London, pp.11-36, 1981.

Fong J.T., Safety Factor on Defect Sizes- A Combined Statistical and Engineering Approach, Material Evaluation, Vol. 39(10), A.14-A.28, 1981.

Forssell C., Ekonomi och Byggnadsvasen, Sunt Fornoft, pp. 74-77, April 1924. (English translation of the excerpts by Lind N. C., in *Structural Reliability and Codified Design* (Lind N. C., ed.), pp. XIII-XVIII, Waterloo University Press, 1970.

Fragola J.R., Reliability Data Bases: The Current Picture, *Hazard Prevention*, 24-29, Jan./Feb. 1987.

Fragola J.R., Reliability and Risk Analysis Data Base Development: An Historical Perpective, *Reliability Engineering and System Safety*, Vol.51, 125-137, 1996.

Frangopol D.M., Progress in Probabilistic Mechanics and Structural Reliability, *Computers and Structures*, Vol. 80(12), 1025-1026, 2002.

Frangopol D.M. and Maute K.; Life Cycle Reliability-Based Optimization of Civil and Aerospace Structures, *Computers and Structures*, Vol. 81, 397-401, 2003.

Freudenthal A.M., Allowable Stresses and Safety of Strutures, *Journal of Association of engineers in Paletine*, 149-153, 1938 (in Hebrew).

Freudenthal A.M., Safety of Structures, *Transactions ASCE*, Vol. 112, pp. 125-180, 1947.

Freudenthal A.M., The Safety Factor, *The Inelastic Behavior of Engineering Materials and Structures*, Wiley, New York, pp. 477-480, 1950.

Freudenthal A.M., Safety and Probability of Structural Failure, *Transactions ASCE*, Vol. 121, 1337-1375, 1956.

Freudenthal A.M., Safety, Reliability and Structural Design, *Journal of the Structural Division*, Vol.87(3),1961.

Freudenthal A.M., Safety, Reliability and Structural Design, *Transactions, ASCE*, Vol. 127, 1962.

Freudenthal A.M., Safety, Safety Factors and Reliability of Mechanical Systems, *Proceedings*, First Symposium on Engineering Applications of Random Function Theory and Probability (Bogdanoff J. L., and Kozin F., eds.) pp.130-162, Wiley, New York, 1963.

Freudenthal A.M., Die Sicherheit der Baukonstruktionen, *Acta Techn. Hung.*, Vol. 46, 417-446, 1964 (in German).

Freudenthal A.M., Critical Appraisal of Safety Criteria and Their Basic Concepts, Preliminary Publication, 8th *Congress, International Association of Bridge and Structural Engineers*, New York, 1968.

Freudenthal A.M., Introductory Remarks, in *"International Conference on Structural Safety and Reliability"*, (Freudenthal A.M., ed.), Pergamon Press, Oxford, pp. 5-6, 1972.

Freudenthal A.M., Basic Theory for Reliability Analysis and Design, in *Reliability problems in Structural Engineering* (Freudenthal A.m. et al, eds.), Maruzen Co.,pp.4-16, Tokyo, 1975.

Freudenthal A.M., Garrelts J.M. and Shinozuka M., The Analysis of Structural Safety, *Journal of Structural Division*, Vol 92, 1966.

Freudenthal A.M. and Schuëller G.I., Risikoanalyze von Ingenieurtragwerken, Reports, *Konstr. Ingenieurbau* (W. Zerna, ed.), Report No. 25, Vulkan-Verlag, Essen, pp.7-95, August 1976 (in German).

Freudenthal A.M., Shinozuka M., Konishi I., And Kanazawa T., *Reliability Approach in Structural Engineering*, Maruzen Co.,Tokyo, 1975.

Frieze P.A., Das P.K and Faulkner D., Partial Safety Factor for Stringer Stiffened Cylinders Under Extreme Compressive Loads, *Report NO. NAOE – 83- 55*,

Department of Naval Architecture and Ocean Engineering, University of Glasgow, U.K., 1983.

Fujino Y. and Lind N.C., Proof Load Factors and Reliability, *Journal of Structural Division*, Vol. 103, 853-870, 1977.

Galambos T.V., Ellingwood B., Macgregor J.G. and Cornell C.A., Probability-Based Load Criteria: Assessment of Current Design Practice, *Journal of Structural Division*, Vol.108, 1982.

Galilei G., *Dialogues Concerning Two New Sciences*, Prometheus Books, Buffalo, NY, 1991 (first published in Italian in 1638).

Garrick B.J., Perspective on the Use of Risk Assessment to Address Terrorism, *Risk Analysis*, Vol.22, 421-423, 2002.

Gaul L., Private Communication, July 18, 2003.

Gemmerling A.V., About the Reliability of Mass Produced Structures (Discussion), Stroitelnaya Mekhanika i Raschet Sooruzhenii, No.5, 1974 (in Russian).

Gemmerling A.V., About Determination of Structural Reliability, *Stroitelnaya Mekhanika i Raschet Sooruzhenii*, No.6.54-57, 1984 (in Russian).

Gertsbakh I. B., *Statistical Reliability Theory*. Marcel Dekker, NewYork, 1989.

Gertsbakh I. B. and Kordonsky Kh. B., Models of Failure, Springer, Berlin, 1969.

Ghare P.M., Quality and Safety Factors in Reliability, *Proc. Ninth Reliability and Maintainability Conference*, Detroit, MI, paper 700655, Vol. 9, 637-641, 1970.

Ghiocel D. and Lungu D., *Safety of Structures*, The Institute of Civil Engineering, Bucharet, 1973 (in Romanian).

Ghiocel D. and Lungu D., *Wind, Snow and Temperature Effects on Structures Based on Probability*, Abacus Press, Turnbridge Wells, Kent, UK, 1975.

Gillmor C.S., *Charles Augustin de Coulomb: Physics and Engineering in Eighteenth Century France*, Princeton University Press, 1971.

Gizzane W., Safety Factor in Car Design, *British Medical Journal*, Vol. 1 (Feb 16), p.399, 1957.

Gladkii V.F., *Strength, Vibration and Reliability of Structures of Flying Vehicles*, "Nauka" Publishing House, Moscow, 1975 (in Russian).

Gnedenko B. V., Belyev Yu. K., and Solovyov A. D., *Mathematical Methods of Reliability Theory*, Academic Press, 1969.

Gnedenko B.V. and Ushakov I.A., *Probabilistic Reliability Engineering*, Wiley, New York, 1995.

Godwin H.J., *J. American Statistic Assoc.,* Vol. 50, 923-945, 1955.

Gofman Y.M., Determining the Heat-Resistance Safety Factor During Long Periods of Operation of 12 MoCr Steel Tubes According to Phase Composition, *Teploenergetica*, Vol.17, 71-73, 1970.

Goldenblatt I.I. and Kopnov V.A., *Criteria of Strength and Plasticity of Structural Materials*, "Mashinostroenie" Publishing House, Moscow, 1968 (in Russian).

Gonvindarajulu Z., Distribution – Free Confidence Bounds for $P(X<Y)$, *Annals of the Inst.*, of Statistical Methods, 229-238, 1968.

Good I.J., Reliability Always Depends on Probability of Course, *Journal of Statistical Computation and Simulation*, Vol.52, 192-193, 1995.

Good I.J., Reply to Prof. Y.Ben-Haim, *Journal of Statistical Computation and Simulation*, Vol. 55, 265-266, 1996.

Gopalan M.N. and Venkateswarlu P., Reliability and Safety Factor of Two-Unit Redundant System, *Reliability Engineering*, Vol.14, 183-192, 1986.

Gore A., The National Information Infrastructure, *Journal of Computers and Science Teaching*, Vol. 14 (1?2), 27-33,1995.

Gorson-Milgrim Sh., Stories of Courage, Mercury Books, Inc., Philadelphia, p.211, 1962.

Gottfried P. and Weiss D.W., Reliability Prediction with Inadequate Data, *Annals of Sixth Reliability and Maintainability Conference*, Coco Beach, FL., 600-607, July 1967.

Govindarajulu Z., Two Sided Confidence Limits for $P(X<Y)$ for Normal Sample of X and Y, *Sankhyā*, Series B, Vol. 29, Part 1 and 2, 35-40, 1967.

Grandhi R.V. and Wang L., Higher-Order Probability Calculation using Nonlinear Approximations, *Computational Methods in Applied Mechanics and Engineering*, Vol.168, 185-206, 1999.

Grandori G., Guagenti E. and Tagliani A., Seismic Hazard Analysis: How to Measure Uncertainty?, *Computers and Structures*, Vol. 67, 47-51, 1998.

Green N. B., Structural Failures: They Point Up Shortcoming in Building Codes, *Western Construction*, Vol. 30(8), 51-52, 1955.

Grigoriu M.D., A Decision Theoretic Approach to Model Selection for Structural Reliability,Ph.D. Thesis M.I.T., Department of Civil Engineering, Cambridge, Mass., 1976.

Grigoriu M. and Turkstra C., Safety of Structures with Correlated Resistances, *Applied Mathematical Modeling*, Vol. 3, pp. 130-136, 1979.

Grimmelt M. J., and Schuëller G. J., Benchmark Study on Methods to Determine Collapse Probabilities of Redundant Structures, *Structural Safety*, Vol. 1, 93-106, 1982.

Gromatskii V.A., About Methods of Establishing Control Loads and Estimating the Structural Reliability by the Experimental Results, *Stroitelnaya Mekhanika i Raschet Sooruzhenii*, No.5.7-14,1984 (in Russian).

Gross-Weege J., On the Numerical Assessment of the Safety Factor of Elastic-Plastic Structures under Variable Loading, *International journal of Mechanical Sciences*, Vol. 39(4), 417-433, 1997.

Gumbel E.J., *Ann. Inst. H. Poincare*, Vol. 5, 115-158, 1935.

Gumbel E. J., *Statistics of Extremes*, Columbia University Press, New York, 1958.

Gumbel, E. J., Bivariate Exponential Distributions, *American Statistical Association Journal*, 1960, pp. 698-707,1960.

Gumbel H., Emil J. Gumbel 1891-1966,Obituary Note, *International Statistical Institute Review*, Vol. 35, 104-105, 1967.

Gupta R.D. and Gupta R.C., Estimation of *P(aTx >bTy)* in the Multivariable Normal Case, *Statistics*, Vol. 21, 91-97, 1990.

Gusev A.S., Towards the Theory of Reliability of Aging Elements, in *Reliability Problems in Structural Mechanics* (V.V. Bolotin and A. Chiras, eds.), RINTIP Publishers, Vilnius 1968 (in Russian).

Guttman I., Bhattacharyya G.K., and Johnson R.A., and Reiser B., Confidence Limit for Stress-Strength Models with Explanatory Variables, *Technometrics*, Vol. 30, 161-168, 1988.

Guttman I., Reiser B., Bhattacharyya GK, Johnson R.A., Bayesian Interference for Stress-Strength Models with Explanatory Variables, in *Gujarat Statistical Review, Professor C.G. Khatri Memorial Volume*, pp. 53-67, 1990.

Hadjian A-H., Issues with Partial Safety Factors, in *Structural Safety and Reliability* (Corotis R.B.,Schuëller G.I.and Shinozuka M., eds.),p.115, Balkema, Lisse, 2001.

Hai-Ning C., Wei-Yang Y., Zhen-Bo H., Wan-Yu C. and Rui-Xiang Z., Analysis of Observation Data of Quanshui Arch Dam and Estimation of its Strength Safety Factor, *Proceedings of the 13th International Congress On Large Dams*, pp.509-526, 1979.

Haimes Y.Y., Barry T. and Lambert J.H.(eds.), When and How Can You Specify a Probability Distribution When You Don't know Much?, *Risk Analysis*, Vol. 14(5), 661-706, 1999.

Haldar A. and Mahadevan S., *Probability, Reliability and Statistical Methods in Engineering Design*, Wiley, New York, pp. 90 and 186, 2000.

Halperin M., Gilbert P.R. and Lachin J.M., Distribution Free Confidence Intervals for P(X1 >X2), *Biometrics*, Vol. 13, 71-80, 1987.

Hamilton C.W. and Drennan J.E., Research Toward a Bayesian Procedure for Calculating System Reliability, pp. 614-620, *Proceedings of Third Aerospace Reliability and Maintainability Conference*, SAE Press, Warrendale, PA, 1964.

Hamilton S.B., Charles Auguste de Coulomb: A Bicentenary Appreciation of a Pioneer in the Science of Construction, *Trans. Newcomen Soc.*, Vol. 17, p. 27, 1936-1937.

Hamilton S.B., Charles Augustin de Coulomb, *Trans. Newcomen Soc. London*, Vol. 17, 27-49, 1938.

Hamilton S.B., The French Civil Engineers of the Eighteenth Century, *Trans. Newcomen Soc.*, Vol. 22, p. 149, 1941-1942.

Harr M.E., *Reliability-Based Design in Civil Engineering*, McGraw-Hill, New York, 1987.

Harr M.E. and Sipher D.J., Reliability and Factor of Safety due to Piping, *Symposium on Water in Mining and Underground Works*, Granads, Spain, 1978.

Harris B and Soms A.P., A Note on Diffulty Inherent in Estimating Reliability From stress-Strength Relationships, *Naval Research Logistiv Quarterly*, Vol. 30, 659-660, 1983.

Hart G. C., *Uncertainty Analysis, Loads, and Safety in Structural Engineering*, Prentice Hall, Inc., Englewood Cliffs, N. J. 1982.

Hartman P., Luetjen P. and Buckberg M., Lower Limits for Total Ship Reliability, *Proceedings of Annual Reliability and Maintainability Symposium*, pp. 210-215, IEEE Press, New York, 1984.

Hasofer A.M., Probabilistic Methods in Structural Engineering, in *Engineering in the Seventies*, The Institution of Engineers, Australia, 1970.

Hasofer A.M., Objective Probabilities for Unique Objects, in Risk, *Structural Engineering and Human Error* (M.Grigoriu, ed.), pp.1-16, University of Waterloo Press, 1984.

Hasofer A. M. and Lind N. C., Exact and Invariant Second-Moment Code format, *Journal of the Engineering Mechanics Division*, Vol. 100, No. EM1, 111-121, 1974.

Haugen E.B., Implementing a Structural Reliability Program, *Proceedings of Eleventh National Symposium on Reliability and Quality Control*, pp 158-168, IEEE Press, New York, 1965.

Haugen E.B., *Probabilistic Approaches to Design*, John Wiley, New York, 1968.

Haugen E. B., *Probabilistic Mechanical Design*, Wiley-Interscience, New York, 1980.

Heffley R.K. and Jewell W.F., Study of Safety Margin System for Powered – Lift STOL Aircraft, *Report No. 1095-1*, May 1978.

Heinfling G., Pendola M., and Hornet P., Reliability Level Provided by Safety Factors, in *Defect Assessment Procedures*, Pressure Vessels Piping Division Publication, Vol. 386, pp. 25-32, 1999.

Heller A.S. and Lelkers E., Reliability Predictions for a Pressurized Water Reactor during the Design Process, *Proceedings of the Annual Reliability and Maintainability Symposium*, pp.348-354, IEEE Press, New York, 1978.

Heller R.A. and Shinozuka M., State-of-the-Art: Reliability Techniques in Materials and Structures, *Proceedings of Ninth Reliability and Maintainability Conference* (Annals of Reliability and Maintainability), Vol. 9, pp. 635-636, SAE Press, Warrendale, PA, 1970.

Heller R.A., Schmidt A. and Deminghoff R., The "Weakest Link" Concept After Proof testing, *Proceedings, IUTAM Symposium on Probabilistic Methods in the Mechanics of Solids and Structures* (Eggwertz S. ed.), Springer Verlag, Berlin, 1984.

Heller R.A., Thangjitham S. and Wall L.L., Probability of Failure of a Proof-Loaded Composite Plate with a Circular Hole, *Proceedings of International Symposium on Composite Materials and Structures*, pp.764-769 Beijing, 1986.

Herrmann C. R., Ingram G. E. and Welker E. L., *An Application of the 'Requirement vs. Capability' Analysis to Estimating Design Reliability of Solid Rocket Motors*, NASA CR-1503, 1970.

Heyman J., *Coulomb's Memoir on Statics*, pp. 20, 54, 85, Cambridge at the University Press, 1972.

Heyman J., The Safety of Masonry Arches, *International Journal of Mechanical Sciences*, Vol. 11, 363, 1969.

Hine J.L., Safety Factor in Car Design, *British Medical Journal*, Vol. 1 (May 11), 1121-1122, 1957.

Hoffman D., and Lewis E.V., Analysis and Interpretation of Full-Scale Data on Midship Bending Stresses of Dry Cargo Ships, *Ship Structure Committee*, SSC-196, U.S. Coast Guard Hqtrs., Washington, D.C., June 1969.

Hollister S.C., The Life and Works of Charles Augustin Coulomb, *Mechanical Engineering*, 615, 1936.

Hong H.P. and Lind N.C, Proof Load Test Levels by Exact Integration, *Canadian Journal of Civil Engineering*, Vol. 18, 297-302, 1991.

Horton W. H., and Durham S. C., Imperfection, a Main Contributor to Scatter in Experimental Values of Buckling Load, *International Journal of Solids and Structures*, Vol.1, 59-72, 1965.

Hoshiya M., Reliability vs. Uncertainty in Structural Safety, *Applications of Statistics and Probability* (Melchers R.E. and Stewart M.G., eds.), Balkema, Rotterdam, Vol. 2, pp. 1131-1134, 2000.

Houlberg W. A., and Baylor L. R., Neoclassical Aspects of Transport in ITER Plasmas with High Axial Safety Factors, *Fusion Technology*, Vol.34(3), Part 2, 591-595, 1998.

Howell G.H., Factors of Safety, *Machine Design*, 76-81, 1956.

Hunsley C. (ed.), *Reliability of Mechanical Systems*, 2nd ed., Mechanical Engineering Publication, London, 1994.

Hutcheon N.B., Safety in Buildings, *Canadian Building Digest*, CBD 114, National Research Council, Division of Building Research, June 1969.

Ibbs C.W., Jr and Crandall K.C., Construction Risk: Multiattriboute Approach, *Journal of Construction Division*, Vol.108, 1982.

Ichikawa M., New Formula for Upper Bound Probability of Failure, *Reliability Engineering*, Vol. 5(3), 173-180, 1983.

Ichikawa M., Confidence Limit of Probability of Failure Based on Stress-Strength Model, *Reliability Engineering*, Vol. 8, 75-83, 1984a.

Ichikawa M., General Formula for Upper Bound of Probability of Failure, *Reliability Engineering*, Vol. 8(1), p. 57, 1984b.

Ichikawa M., Proposal of a New Distribution-Free Method for Reliability Demonstration Tests, *Reliability Engineering*, Vol.9, 99-105, 1984c.

Ichikawa M., Some Complementary Notes to the Distribution-Free Method for Reliability Demonstration, *Reliability Engineering*, Vol. 16, 311-315, 1986.

Ichikava M., Improved Formula for Upper Bound of Probability of Failure Based on Means and Variances, in *Probabilistic Structural Mechanics: Advances in Structural Reliability Methods* (Spanos P.D. and Wu Y.-T., eds.), Springer, Berlin, pp.293-300, 1994.

Ichikava M. and Okabe N., Some Problems in Probabilistic Fracture Mechanics, in *Recent Studies on Structural Safety and Reliability* (Nakagawa T., Ishikawa H. and Tsurui A., eds.), Elsevier Applied Science, London, pp. 1-24, 1989.

Iliasevich C.A., N.S. Streletskii – Founder of the Soviet School of Metallbuilding, in *Selected Works* by N.S. Streletskii (Belenia E.I., ed.), pp. 8-26, "Stroiizdat" Publishers, Moscow, 1975 (in Russian).

Ilyuksne N.I. and Gonchavov V.P., Safety Factor Determination During Cyclic Loading taking into Account of Intial Stresses, *Problemy Prochnosti*, No.2, 29-32, 1979 (in Russian)

Ingles O.G., Safety in Civil Engineering – Its Perception and Promotion, *Reliability Engineering*, Vol. 1, 15-27, 1980.

Ingram G.E. and Herrman C.R., Structural Reliability Methodology – Application at the System Level, *Nuclear Engineering and Design*, Vol. 50, 185-193, 1978.

Innis C.L. and Hammand T., Predicting Mechanical Design Reliability using Weighted Fault Trees, *Failure Prevention and Reliability*, pp. 213-228, ASME Press, New York, 1977.

Ismail R., Jeyratnam S. and Panchapakesan S., Estimation of Pr[X>Y] for Gamma Distributions, *Statist. Computation and Simulation*, Vol. 26, 253-267, 1986.

Iuculano G. and Zanini A., Evaluation of Failure Models through Step-Stress Tests, *IEEE Transactions on Reliability*, Vol. R-35, No.4, 409-413, 1986.

Ivshin V.V. and Lumelskii Y.P., *Statistical Problems of Estimation on the Model "Load-Strength"*, Perm University Press, Perm, 1995 (in Russian).

Jensen F., It Mustn't Break (editorial), *Quality and Reliability Engineering International*, Vol.16, p.1, 2000.

Johnson A. I., *Strength, Safety and Economical Dimension of Structures*, Bulletin No. 12, Royal Institute of Technology, Stockholm, 1953 (second edition 1971).

Johnson M. and Matthews F.L, Determination of Safety Factor, for Use when Designing Bolted Joints in GRP, *Composites*, 73-76, April 1979.

Johnson N.L., and Kotz S., *Leading Personalities in Statistical Sciences*, Wiley-Interscience, New York, 1997.

226

Johnson R.A., Stress-Strength Models for Reliability, *Handbook for Statistics*, Vol. 7 (Krishnaiah P.R. and Rao R., eds.), North Holland, 1988.

Jones F., Safety Factor in Car Design, *British Medical Journal*, Vol. 1 (April 6), p.820, 1957.

Jones J.L. and Jones R.E., Determination of Safety Factor for Defibrillator Waveforms in Cultured Heart Cells, *American Journal of Physiology*, Vol. 242(4), H.662-H.670, 1982.

Jouris G.M and Shaffer D.H., Methodology for Quantitative Assembly of the Impact of Safety Factors on Structural integrity, in *Structural Integrity Technology* (Gallagher J.P. and Crooker T.W., eds.), ASME Press, pp. 45-47, 1979.

Jovitt P.W., Bayesian Estimates of Material Properties from Limited Test Data, *Engineering Structures*, Vol. 1, 170-178, 1979.

Julian O.G., Synopsis of First Progress Report of Committee on Factors of Safety, *Journal of Structural Division*, 1316.1-1316.22, July 1957.

Juvinal R.C., *Fundamentals of Machine Component Design*, John Wiley, New York, 1983.

Kafka P., Some Thoughts on the Interaction between System and Structural Reliability, *Nuclear Engineering and Design*, Vol. 60, 129-132, 1980.

Kaliszky S., Gábor Kazinczy, *Periodica Polytechnica*, Civil Engineering, Vol. 28 (1-4), 75-93, 1984.

Kaliszky S., Private Communication, July 8, 2003

Kalmins A. and Updike D.P., Safety Margins if Pressure Vessels when Designed By Primary Stress Limits, *ASME Pressure Vessels and Piping Division Publication*, PVP Vol. 277, pp. 79-88, ASME Press, New York, 1994.

Kameda M. and Koike T., Reliability Theory of Deteriorating Structures, *Journal of Structural Division,* Vol. 101, 295-310, 1975.

Kamenjarzh J., *Limit Analysis of Solids and Structures*, CRC Press, Boca Raton, 1996.

Kamenjarzh J. and Weichert D., On the Kinemstic Upper-Bounds for Safety Factors in Shakedown Theory, *International Journal of Plasticity*, Vol. 8(7), 827-837, 1992.

Kamiyama T., Structureal Reliability and Safety Factors, *Redstone Scientific Information Center Report RS1C-463*, Sept. 1965.

Kanda J., Normalized Failure Cost as a Measure of a Structure Importance, *Nuclear Engineering and Design*, Vol. 160, 299-305, 1996.

Kanda J. and Ellingwood B., Formulation of Load Factors Based on Optimum Reliability, *Structural Safety*, Vol. 9, 197-210, 1991.

Kao J.H.K., Statistical Models in Mechanical Reliability, *Proceedings of the National*

Kaplan S.M., Factors of Safety for Unmanned Spacecraft Structures, *Shock and Vibration Bulletin*, Vol. 40, 89-97.

Kapur K.C., Reliability Bounds in Probabilistic Design, IEEE Transactions on Reliability, Vol. R-24(3), 193-195, 1975.

Kapur K.C., Techniques of Estimating Reliability at Design Stage, in *Handbook of Reliability, Engineering and Management* (Ireson W.G. and Coombs C.F., eds), McGraw-Hill, New York, Chapter 18, 1988.

Kapur K.C. and Lamberson L.R., *Reliability in Engineering Design*, Wiley, New York, 1977.

Kapur K.C. and Lamberson L.R, Inference Theory and Reliability Computations, in *Reliability in Engineering Design*, Marcel Dekler, New York, 1983.

Kanzinczy G., Beitrag zum Vortrag Gehlers, *Internationale Tagung für Brückenbau und Hochbau*, Schlussbericht, Wien, 1928 (in German).

Kazinczy G., Zulässigkeit der Anwendung des Traglastverfahrens bei Stahlbeton, *Internationale Vereinigung für Brückenbau*, p. 135, Schlussbericht, London, 1952.

Kececioglu D., *Reliability Engineering Handbook*, Vol.1, Prentice Hall, Englewood Cliffs, New Jersey, 1991.

Kecicioglu D., Reliability Analysis of Mechanical Components and Systems, *Nuclear Engineering and Design*, Vol. 19, 259-290, 1972.

Kececioglu D. and Haugen E.B., A Unified Look at Design Safety Factors, Safety Margins and Measures of Reliability, in *Annals of Assurance Sciences-Seventh Reliability and Maintainability Conference*, pp. 520-530, 1968.

Kececioglu D. and Li D., Exact Solutions for the Prediction of the Reliability of Mechanical Components and Structural Members, *Proceedings of the Failure Prevention and Reliability Conference*, pp. 115-122, ASME Press, New York, 1985.

Kececioglu D., Why Design by Reliability?, *Proceedings of Seventh Reliability and Maintainability Conference* (Annals of Assurance Sciences, Vol. 7), p. 491, ASME Press, New York, 1968.

Kecicioglu D. and Li D., Aspect of Unreliability and Reliability Determination by Stress/Strength Interference Approach, in *Probabilistic Structural Analysis* (Sundararajan C.(Raj)), PVP-Vol.93, ASME Press, New York, pp.75-100, 1984.

Kececioglu D. and Cormier D., Designing a Specified Reliability Directly into a Component, in *Proceedings of the Third Annual Aerospace Reliability and Maintability Conference*, Washington D.C., pp.546-565, 1967.

Kececioglu D. and Haugen E.B., A Unified Look at Design Safety Factors, Safety Margins, and Measures of Reliability, *Proceedings of Seventh Reliability and Maintainability Conference (Annals of Assurance Sciences, Vol. 7)*, pp. 520-528, ASME Press, New York, 1968.

Keith-Lucas D., Design for Safety, *Aeronautical Journal*, Vol. 77, 483-488, 1973.

Kellerman O., Some Aspects of Probabilistic Structural Reliability to Meet the Needs of Risk Analysis of Nuclear Power Plants, *Nuclear Engineering and Design*, Vol. 60, 157-158, 1980.

228

Kezdi A., Safety Factors for Different Types of Failures, *Proceedings of the 7ᵗʰ European Conference on Soil Mechanics and Foundation Engineering*, 195-198, Sept. 1979.

Kharionovskii V.V., Stochastic Methods in Calculations of Cross-Country Pipelines, *Mechanics of Solids*, Vol. 31(3), 91-96, 1996.

Khinchin A.I., *Mathematical Foundations of Information Theory*, Dover Publications, New York, 1957.

Khozialov, N.F., Safety Factors, *Building Ind.*, Vol. 10, 840-844, 1929 (in Russian).

Kirkpatrick I. Predicting Reliability of Electromechanical Devices, *Proceedings of Sixth National Symposium on Reliability and Quality Control*, pp. 272-281, IEEE Press, New York, 1960.

Klein M., Personal communication, September 2, 1998.

Klein M., Probabilistic Approach to Structural Factor of Safety, Delft, The Netherlands March 27, 1995 (personal communication dated Sept. 2, 1998).

Klein M., Schuëller G.I., Deymarie P., Macke M., Courrian P. and Capitanio R.S., Probabilistic Approach to Structural Factors of Safety in Aerospace, in Proceedings, *International Conference on Spacecraft Structures and Mechanical Testing*, Paris, France, 1999.

Klein W.A. and Lehr S.N., Reliability of Solar Arrays, IRE *Transactions on Reliability and Quality Control*, pp. 71-80, Vol. RQC-11, No. 3, 1962.

Knell F., Commentary on the Basic Philosophy and Recent Development of Safety Margins, *Canadian Journal of Civil Engineering*, Vol.3, 409-416, 1976.

Knowles J., Engineering Reliability into Today's Automotive Vehicles, *Proceedings of the Annual Symposium on Reliability*, pp. 64-74, 1968.

Kobayashi A.S. (ed.), *Manual of Engineering Stress Analysis*, Prentice-Hall, Englewood Cliffs, NJ, 1982.

Kogan J., *Lifting and Conveying Machinery*, Iowa University Press, Ames, p.186, 1985.

Koiter W.T., Fail-Safe Structural Design (discussion), *Journal of the Royal Aeronautical Society*, Vol. 62, No. 574, 1958.

Kokhanenko I. K., Statistical Method of Determining the Safety Factor, *Izvestiya Vuzov, Mashinostroenie*, No. 12, 1983, pp. 9-12 (in Russian).

Kolosov L.V and Dzhuar V.V., The Life and Safety Factors of Rope Made of Rubber-Embedded Steel Cables, *Soviet Engineering Research*, Vol. 4, 42-44, 1985.

Komar N.M. and Okopnii Yu.A., An Experimental Check of the Reliability Function's Estimate on Electronic Models, in *Reliability Problems in Structural Mechanics*, (V.V. Bolotin and A.A. Chiras, eds.), pp. 69-74, "RINTIP" Publishers, Vilnius, 1971 (in Russian).

König G. and Heunisch M., *Zur Statistischen Sicherheitstheorie in Stahlbetonbau*, Verlag von Wilhelm Ernst&Sohn, Berlin, 1972 (in German).

Konishi I.., Structural Reliability Analysis Considering Strength Assurance Level , in *Reliability Problems in Structural Engineering* (Freudenthal A.M. et al, eds.), Maruzen Co.,pp.91-110, Tokyo, 1975.

Korbashov V.F., Unified Criteria for Calculating the Safety Factor of Slope Stability for Earth Dams, *Hydrotechnical Construction*, No.7, 1190-1194, 1978.

Korol B. and Lang W.J., Determination of Physical Safety Factor of Potential Pharmacological Agents, *Journal of Pharmacy Science*, Vol. 54(10), p.1555, 1965.

Koslov B. and Ushakov I. A., *Handbook on Reliability Computation*, "Sovietskoe Radio" Publishers, Moscow, 1975 (in Russian).

Kotz S., Lumelskii Y. and Pensky M., *The Stress-Strength Model and Its Generalizations*, World Scientific, Singapore, 2002.

Kozlov B.A. and Ushakov I.A., *Reliability Handbook*, Holt, Rinehart and Winston, New York, 1970.

Kozoriz E.P., Planetary Quarantine and Its Challenge to Materials Reliability, *Proceedings on Sixth Reliability and Maintainability Conference (Annals of Reliability and Maintainability, Vol. 6)*, pp. 395-404, SAE press, Warrendale, PA, 1967.

Kragelskii I.V. and Schedrov V.S., *Development of Science of Friction – Dry Friction*, pp. 51-69, Moscow, 1956.

Kranakis E.F., Navier: Theory of Suspension Bridges, From Ancient Omens to Statistical Mechanics, *Acta Hist. Sci. Nat. Med.*, Vol. 39, 247-258, 1987.

Kreuzer H.L., A Probability Based Safety Factor Approach for Arch Dams, *Water Power*, 458-463, Dec. 1973.

Krinitzsky E., Earthquake Probability in Engineering, Part 1: The Use and Misuse of Expert Opinion, *Engineering Geology*, Vol. 33, 257-288, 1993.

Kürschners Deutscher Gelehrten-Kalender, 11th ed., p. 3427, Berlin, 1971 (in German).

Kürschners Deutscher Gelehrten-Kalender, 3rd ed., p. 1516, 1928/1929 (in German).

Kürschners Deutscher Gelehrten-Kalender, 7th ed., p. 1311, 1950 (in German).

Kürschners Deutscher Gelehrten-Kalender, 9th ed., p. 1294, 1961 (in German).

Kullbeck S., *Information Theory and Statistics*, Dover, New York, 1959.

Kunreuther H., Risk Analysis and Risk Management in an Uncertain World, *Risk Analysis*, Vol.22, 655-664, 2002.

Kurtz P.H., A Strength-Stress Interference Model for Time-Varying Probability of Failure and Its Applications to Electromechanical System, Naval Sea Systems Command, Rept. TM-75-122, 1975 (available from NTIS as AD-A016299).

Kuznetsov A.A., *Design Reliability of Ballistic Missiles*, Moscow, 1978 (in Russian).

Kuznetsov A.A., Alifanov O.M., Vetrov V.I., Zolotov A.A. and Titov M.I., *Probabilistic Characteristic of Strength of Aviation Materials and Dimentions*, A Handbook, "Mashinostroevnie" Publisher, Moscow, 1970 (in Russian).

230

Langejan A., Some Aspect of the Safety Factor in Soil Mechanics Considered as a Problem of Probability, Proceedings of the 6th International Conference on Soil Mechanics and Foundations Engineering, Montreal, Vol.2, 1965.

Langejan A., Some Aspects of the Safety Factor in Soil Mechanics Considered as a Problem of Probability, *Proc. 6th Int. Conf Soil Mechanics and Foundation Engineering*, Vol.2, Montreal, Canada, 1965.

Lapple C.F., Safety Factor, *Aviation Week and Space Technology*, Vol. 120(8), p. 97, 1984.

Larabee R. and Cornell C.A. (eds.), *Probabilistic Mechanics and Structural Reliability*, ASCE Press, New York, 1979.

Lebris P, Three Safety Factors Included in LNG Storage Tank System, *Pipe Line Industry*, 45-50, Sept.1978.

Lee J.H. and San S.Y., Reliability Consistent Load and Resistant Factors, in *Proceedings, 5th International Conference on Structural Safety and Reliability*, (Ang A.H-S., Shinozuka M. and Schuëller G.I., eds.), pp.2365-2369, 1989.

Lemon G.H. and Manning S.D., Literature Survey on Structural Reliability, *IEEE Transactions on Reliability*, Vol. R-23, No. 4, 263-366, 1974.

Lemon G.H. and Manning S.D., Literature Survey on Structureal Reliability, *IEEE Transactions on Reliability*, Vol. R-23(4), 263-266, 1974.

Leporati E., *The Assesment of Structural Safety*, Research Studies Press, Forest Groves, Oregon, 1977.

Le Seur T., Jacquier F. and Bosovich R.G., Parere di tre mattematici sopra i danni che si sono trovati nella cupola di S. Pietro sul fine dell' Anno 1742, Rome, 1743 (in Latin).

Leunberger D.G., *Introduction to Linear and Non-linear Programming*, Addison-Wesley, Reading, MA, 1984.

Levi R.,Calculus Probabilistes de la Securite des Constructions, *Ann. Ponts et Chaussees*, Vol. 119, No. 4, 493-539, 1949.(In French)

Lewis E.E., *Introduction to Reliability Engineering*, Wiley, New York, p171,1987.

Li H. and Foschi O., An Inverse Reliability Method and Its Application, *Structural Safety*, Vol.20, 257-270, 1998

Libertiny G. Z., Safe Design without Safety Factors, *The 1997 ASME International Mechanical Engineering Congress and Exposition*, Dallas, Texas, Nov.16 – Nov.21, 1997, American Society of Mechanical Engineering PAP, 1997.

Lie T.T., Safety Factors for Fire Loads, *Canadian Journal of Civil Engineering*, Vol. 6, 617-628, 1979.

Liebowitz H., Navy Reliability Research, *Proceedings of Sixth Reliability and Maintainability Conference (Annals of Reliability and Maintainability, Vol. 6)*, pp.34-53, SAE Press, Warrendale, PA, 1967.

Liebowitz H., Preface, *Engineering Fracture Mechanics*, Vol. 8, 5-7, 1976.

Ligtenberg F.K., Structural Safety and Catastrophic Events, in *Proc IABSE (IVBH) Symposium and Concepts of Structures and Methods of Design*, London/Zürich, pp.1-41, 1969.

Lin T.S. and Cheng K.S., Characterization of the Relationship between Motor End-Plate Jitter and the Safety Factor, *Muscle Nerve*, Vol. 21(5), 628-636, 1998.

Lin T.S. and Nowak A.S., Proof Loading and Structural Reliability, *Reliability Engineering*, Vol. 8, 85-100, 1984.

Lind N., Mechanics, Reliability and Society, in *Proceedings of the Canadian Conference in Mechanics*, 1976.

Lind N.C., The Design of Structural Design Norms, *Journal of Structural Mechanics*, Vol. 1, 357-370, 1973.

Lind N.C., Approximate Analysis and Economics of Structures, *Journal of Structural Division*, Vol. 102, 1177-1196, 1976.

Lind N.C., On the Value of Life and Limb, in DIALOG. 4-78, DIAB, *Danish Engineering Academy*, Lyngby, Denmark, Sept. 1978.

Lind N. S., Consistent Partial Safety Factors, *Journal of the Structural Division*, Vol.97, 1651-1669, 1971.

Lind N., Safety Management and Social Progress, in *Sixth International Conference on Applications of Statistics and Probability in Civil Engineering*, (Esteva L., and Ruiz S.E., eds.), Vol.3,pp.17-29, Mexico City, 1991.

Lind N. C., A Measure of Vulnerability and Damage Tolerance, *Reliability Engineering and System Safety*, Vol.48, 1-6, 1995.

Lind N. C., Vulnerability of Damage-Accumulating Systems, *Reliability Engineering and System Safety*, Vol.53, 217-219, 1996a.

Lind N. C., Vulnerability of Multiple Barrier Systems, *Nuclear Safety*, Vol.37, 139-148, 1996b.

Lind N., Private Communication, July 21, 2003.

Lind N.C., Knab L.I. and Hall W.B., Economic Study of the Connection Safety Factor, *Third International Specialty Conference on Cold-Formed Steel Structures*, St. Louis, MO., Nov 24-25, 1975.

Lipson C., New Concepts on Safety Factors, *Product Engineering*, Vol.31275-278, 1960.

Lipson C., and Narendra Sh., Prediction of Percent of Failures for Stress/Strength Interactions, sponsored by RADC, U.S. Air Force, Contract No. AF30 (602) 3684 (quoted in the paper by My Dao Thien and Massoud, 1974).

Lipson C., Sheth N.J., Disney R.L. and Altum M, Reliability Prediction – Mechanical Stress/Strength Interference, *Technical Report No. RADC-TR68-403*, Rome Development Center, New York, 1969.

Lipson C. and Sheth N.Y., *Statistical Design and Analysis of Engineering Experiments*, McGraw Hill Kogakusha, Tokyo,1973.

Lipson C.,Sheth N.Y. and Disney R.L., Reliability Prediction-Mechanical Stress/Strength Interference, *Technical Report No. RAD (-TR-66-710)*, Rome Air Development Center, New York, 1967.

Little R.E., Factor of Safety, *Machine Design*, Vol.38, 165-170, July 1966.

Liu S.C. and Neghabat F., A Cost Optimization Model for Seismic Design of Structures, *The Bell Systems Technical Journal*, Vol.51, 2209-2225, 1972.

Lloyd D. K., and Lipow M., *Reliability: Management, Methods, and Mathematics*, Prentice-Hall, Englewood Cliffs, New Jersey, 1962.

Loll V., Load-Strength Modelling of Mechanics and Electronics, Quality and Reliability Engineering International, Vol.3, 149-155, 1987.

Losaberidze A., In Conjunction of the 80th Birth Anniversary of A. Kakushadze, *Metsniereba da Tekhnika*, No. 4, p. 59, 1984 (in Georgian).

Lovelace A.M., Keynote Address, in *International Conference on Structural Safety and Reliability*, (Freudenthal A.M., ed.) Pergamon Press, Oxford, p. 4, 1972.

Lumb P., Safety Factors and the Probability Distribution of Strength, *Canadian Geotechnical Journal*, Vol.7, 225-242, 1970.

Lundstro L.C., Safety Factor in Automotive Design, *SAE Prog. Technology*, Vol. 13, p. 511, 1968.

Lundstro L.C., Safety Factor in Automotive Design, *SAE Transaction*, Vol. 75, p. 124, 1967.

Lusser R., Reliability through Safety Margins, U.S. Army Ordinance Missile Command, *Redstone Arsenal*, United States Department of Defence, Huntsville, Alabama, Oct.1958.

MacGregor J.G., Safety and Limited States Design for Reinforced Concrete, *Canadian Journal of Civil Engineering*, Vol.3, 484-513, 1976.

MacGregor J.G., Load and Resistance Factors in Concrete Design, *ACI Journal*, 279-287, July-August 1983.

MacGregor J.G., Safety and Limit States Design for Reinforced Concrete, *Canadian Journal of Civil Engineering*, Vol. 3, 484-513, 1976.

McGraw-Hill Dictionary of Engineers, 2nd edition, p.209, McGraw-Hill, New York, 2002.

Mackey A.C. and Larsen C.E., Space Station Freedom. Requirements for Structural Safety, Reliability and Verification, *Proceedings, 5th International Conference on Structural Safety and Reliability* (Ang A.H.-S., Shinozuka M. and Schuëller G.I., eds.), pp.2315-2320, ASCE Press, New York, 1989.

Madsen H.O., Omission Sensitivity Factors, *Structural Safety*, Vol. 5, 35-45, 1988.

Madsen H.O. and Egeland Th., Structural Reliability – Models and Applications, *International Statistical Review*, Vol. 57(3), 185-203, 1989.

Madsen H.O., Krenk S. and Lind N.C., *Methods of Structural Safety*, Prentice Hall, Englewood Cliffs, 1986.

Maes M.A., The Influence of Uncertainties on the Selection of Extreme Values of Environmental Loads and Events, *Civil Engineering Systems*, Vol.7, 115-124, 1990.

Maes M.A. and Huyse L., Tail Effects of Uncertainty Modeling in Risk and Reliability Analysis, *Proceedings of the Third International Symposium on Uncertainty Modeling and Analysis*, College Park, Md., pp.133-138, 1995.

Maes M. A. and Huyse L., Developing Structural Design Criteria with Specified Response Reliability, *Canadian Journal of Civil Engineering*, Vol.24, 201-210, 1997.

Makris S.L., The FQPA 10X Safety Factor: How Traditional Risk Assessment Practices Play a Role in Its Application, *Human Ecology and Risk Assessment*, Vol. 5(5) 1003-1012, 1999.

Mallagh C., The Inherent Unreliability of Reliability Data, *Quality and Reliability Engineering International*, Vol. 4, 35-39, 1988. Management Science, Vol. 18, 454, 1972.

Mangurian G.N., The Aircraft Structural Factor of Safety, *AGARD Report 154*, 1957.

Mann H.B. and Whitney D.R., On a Test of Whether One of Two Random Variables is Stochastically Larger Than the Other, *Ann. Math. Statist.*, Vol. 18, 50-60, 1947.

Markov A.A., *Ischislenie Veroiatnostei*, Gosizdat, Moscow, 1913 (in Russian).

Matsuo M., and Asaoka A., A Statistical Study on Conventional Safety Factor Method, *Soils and Foundations*, Vol.16, 75-90, 1976.

Matsuo M., and Demura Y., Updating the Safety Factor in Structural Design Specifications, *Transactions of Japan Society of Civil Engineers*, Vol.15, 206-208, 1983.

Mau S.T. and Sexsmith R.G., Minimum Expected Cost Estimation, *Journal of Structural Division*, Vol. 98 (9), 2043-2058, 1972.

Mayer, M., *Die Sicherheit der Bauwerke and Ihre Berechnung nach Grenzkräften anstatt nach Zulässigen Spannungen*, Springer, Berlin, 1926 (in German) (see also paper by Tichý, M., Max Mayer --Begründer der Berechnungsmethode nach Grenzzuständen, Technische Mechanik, Magdeburg, Federal Republic of Germany, 1990).

Mayers R.E., How Meaningful are Reliability Predictions, in Proceedings, *1969 Annual Symposium on Reliability*, pp.412-416, 1969.

Maymon G., *Some Engineering Applications in Random Vibrations and Random Structures*, AIAA Press, Reston, VA, 1998.

Maymon G., The Stochastic Safety Factor – A Bridge Between Deterministic and Probabilistic Structural Analyses, *PSAM5*, 2000.

Maymon G., The Stochastic Factor of Safety - A Different Approach to Structural Design, *CD-Rom, Proceedings of the 42nd Israel Annual Conference on Aerospace Sciences*, February 20-21, 2002a.

Maymon G., Personal Communication, September 26, 2002b.

Mazumdar M., Some Estimates of Reliability using Interference Theory, *Naval Research Logistic Quarterly*, Vol.17, 159-165, 1970.

Mazumdar M., Marshall J. A. and Chay S. C., Propagation of Uncertainties in Problems of Structural Reliability, *Nuclear Engineering and Design*, Vol. 50, 163-167, 1978.

McAdams P., Reliability: Continuing challenge in Construction Machinery, *SAE Journal*, Vol. 74, 87-89, 1966.

McCalley R.B., Nomogram for Selection of Safety Factors, *Design News*, 138-141, Sept.1957.

McCartney L.N., Can Safety Factor be Reduced Safety when Designing against Fatigue?, *Fatigue of Engineering Material and Structures* ,Vol. 2, 387-400, 1980.

McCool J.I., Inference P(X>Y) in the Weibull Case, *Communications in Statics*, B: Simulation and Computation, Vol. 20, 129-148, 1991.

McGraw-Hill Dictionary of Scientific and Technical terms, New York, 1975.

Meidell B., Sur La probabilite des erreurs, *Comptes Rendus*, Paris, Vol. 176, 280-282, 1923.

Melchers R.E., Human Error in Structural Reliability Assessments, *Reliability Engineering*, Vol. 7, 61-75, 1984.

Melchers, R. E., *Structural Reliability and Predictions*, Ellis Horwood, London, 1987.

Melchers, R. E., *Structural Reliability Analysis and Predictions*, Wiley, Chichester, p. 23, 1999.

Melloy B.J. and Cavalier T.M., Bounds for the Probability of Failure Resulting from Stress/Strength Interference, *IEEE Transactions on Reliability*, Vol. 38(3), 383-385, 1989.

Mendellson K., A Scientist Looks at the Pyramids, *American Scientist*, 210-220,1971.

Mendelson A., *Plasticity: Theory and Application*, pp. 300-326, Robert E. Krieger Publishing Company, Malabar, FL, 1968.

Mettam J.D. and Berry J.G., Factors of Safety for the Design of Breckwsters, Vol. 63, No. 747, 319- 322, 1983.

Meyerhof G.C., Safety Factors in Soil Mechanics, *Canadian Geotechnical Journal*, Vol.7(4), 1970.

Meyerhof G.C., Safety Factors and Limit States in Geotechnical Engineering, *Canadian Geotechnical Journal*, Vol.21, 1984.

Meyerhof G.G., Safety Factor in Soil Mechanics, *Canadian Geotechnical Journal*, Vol. 21,1-7, 1984.

Millers I. and Freund J. E., Tolerance Limits, in *Probability and Statistics for Engineers*, Prentice hall, Englewood Cliffs, pp. 514-517, 1977.

Millwater H. et al, Recent Developments of the NESSUS Probabilistic Analysis Computer Program, *Proceedings of the 33rd AAIA/ASME/ASCE/AHS/ASC Structures, Structural Dynamics and Materials Conference*, Washington DC, 1992.

Ministerio Das Obras Publicas, Laboratorio Nacional De Engenharia Civil, Lisboa, Portugal, June 1960.

Mischke C., A Method of Relating Factor of Safety and Reliability, *Journal of Engineering for Industry*, Vol B92(3), 537-542, 1970.

Mischke C.R., Some Tentative Weibullean Descriptions of the Properties of Steels, Aluminums, and Titaniums, *ASME Paper 71-Vibr-64*, Sept. 1971.

Mitroff I.I., On The Social Psychology of the Safety Factor: A Case Study in the Sociology of Engineering Science

Mittenberg A.A., Fundamental Aspects of Mechanical Reliability, *Mechanical Reliability Concepts*, pp. 17-34, ASME Press, New York, 1965.

Mochio T., A Proposal for Safety Factor Estimation in Seismic PSA of Structure by Stochastic FEM Technique, *Nuclear Engineering and Design*, Vol. 147(2) 129-139, 1994.

Modarres M., *What Every Engineer Should Know about Reliability and Risk Analysis*, Marcel Dekker, Inc., New York, p.288, 1993.

Morozov N.P., Reliability in Engineering, *Mashinovederie*, No. 3, 62-67, 1981, (in Russian).

Moses F., Probabilistic-Based Structural Specifications, *Risk Analysis*, Vol. 18(4), 445-454, 1998.

Mukhadze G. M., *Selected Works*, Metsniereba Publishing House, Tbilisi, 1986 (Parts in Georgian; Parts in Russian).

Mukhadze G. and Kakushadze A., *Determination of the Safety Factors by the Methods of Mathematical Statistics*, "Tekhnika da Shroma" Publishers, Tbilisi, 1954 (in Georgian).

Mukhadze L., Private Communication, July 24, 2003.

Mukhadze L., Private Communication, August 18, 2003.

Murata T., Reliability Case History of an Airborne Air Data Computer, *IEEE Transactions on Reliability*, Vol. R-24, No. 2, 98-102, 1975.

Murata T., Reliability Case History of an Airborne Air Turbine Starter, *IEEE Trasactions on Reliability*, Vol. R-25, No. 5, 302-303, 1976.

Murotsu Y. et al, New Method of Relating Safety Factor to Failure Probability in Structural Design, ASME Winter Annual Meeting, California. 1978.

Murotsu Y., New Method of Relating Safety Factor to Failure Probability in Structural Design, in *Advances in Reliability and Stress Analysis* (Burns Y. Y., Jr., ed.), ASME Press, New York, pp. 23-33, 1979.

Murotsu Y., Okada H. Taguchi K., Grimmelt M. and Yonerawa M., Ontematic Generation of Stochastically Dominant Failure Modes of Frame Structures, *Structural Safety*, Vol. 2, 17-25, 1984.

Murzewski J., *The Introduction to the Structural Safety Theory*, Panstwowe Wydawnictwo Naukowe, Warsaw, 1963 (in Polish).

Murzewski, J., *Bezpieczenstwo Konstrukeji Budowlanych*, "Arkady" Publishing House, Warsaw, 1970 (in Polish).

Murzewski J.W., Distribution – Based Level-2 Design, in *Fourth International Conference on Applications of Statistics and Probability in Soil and Structural Engineering*, pp.585-596, Pitagora Editrice, 1983.

Murzewski, J., *Niezawodnosc Konstrukci Inzynierskich*, "Arkady" Publishing, Warsawa, 1989 (in Polish).

My D. T. and Massoud M., On the Relation between the Factor of Safety and Reliability, *Journal of Engineering for Industry*, Vol. 96, 853-857, 1974.

My D. T. and Massoud M., On the Probabilistic Distributions of Stress and Strength in Design Problems, *Journal of Engineering for Industry*, Vol. 97 (3), 986-993, 1975.

Naresky J.J., Reliability Definitions, *IEEE Transitions in Reliability*, Vol. 19, 198-200,

Natke H.G. and Cempel C., *Model-Aided Diagnosis of Mechanical Systems*, Springer, New York, 1997.

Nathwani J., Risk Management Is about Lives (Gaines) versus Lives (Lost)- Not Dollars, *Risk Abstracts*, Vol.7(4), 1-5, 1990.

Nakagiri S. and Hisada ., *An Introduction to Stochastic Finite Element Method: Analysis of Uncertain Structures*, Baifukan, Tokyo, 1985 (in Japanese).

Navier, *Memoire sur les lois de l'équilibre et du movement des corps solides élastiques*, Memoires … de l'Institut, Vol.VII, 1827 (in French).

Neal D. M., Mattew W. T. and Vangel M. G., Model Sensitivities in Stress-Strength Reliability Computations, *Report, Materials Technology Laboratory*, TR 91-3, Watertown, MA, Jan. 1991.

Neogy R., Reliability for N/C Machines – A Must, *Proceedings of Annual Symposium on Reliability*, pp. 101-105, IEEE Press, New York, 1970. New York, 1983.

Norton R. L., *Machine Design*, Prentice Hall, Upper Saddle River, NJ, pp. 19-22, 2000.

Nowak A. S., Risk Analysis for Code Calibration, *Structural Safety*, Vol. 1, 289-304, 1983.

Nowak A. S. and Collins K. R., *Reliability of Strucutres*, McGraw Hill, Boston, 2000.

Nowak A., Private Communication, July 18, 2003.

Nowak A.S. and Szerszen M.M., Calibration of Design Code for Buildings (ACI 318): Part 1 – Statistical Models of Resistance, *ACI Structural Journal*, Vol.100, 377-382, 2003.

Nucci E.Y., Concepts and Research Needs of Reliability in Military Systems, *IEEE Transactions on Reliability*, Vol. R-12, No. 3, 1-4, 1963.

O'Neill R., Future Developments of Probabilistic Structural Reliability, *Nuclear Engineering and Design*, Vol. 60, 151-156, 1980.

Obukhov A.S., Safety Factor of Chemical Apparatus Made of Fiberglass-Plastic, *Chemical Petroleum Engineering*, 761-764, 1969.

O'Conner J.J. and Robertson E.F., Claude Louis Marie Henri Navier, http://www-history.mcs.st-andrews.ac.uk/Mathematicians/Navier.html

O'Conner J.J. and Robertson E.F., Charles Augustin de Coulomb, http://www-history.mcs.st-andrews.ac.uk/Mathematicians/Coulomb.html

O'Connor R.D.T., *Practical Reliability Engineering*, Wiley, New York, 1985.

Oden J.T., Belytschko T., Babuška I. and Hughes T.J.R., Research Directions in Computational Mechanics, *Computer Methods in Applied Mechanics and Engineering*, Vol.192, 913-922, 2003.

Oestergsard C., Partial Safety Factors for Vertical Bending loads on Contareships, *Jornal of Offshore Mechanics and Arctic Engineering*, Vol. 114, 129-136, 1992.

Okubo S., Fukui K. and Nishimatsu Y., Local Safety Factor Applicable to Wide Range of Failure Criteria, *Rock Mechanics and Rock Engineering*, Vol. 30(4), 223-227, 1997.

Olszak W., Kaufman S., Elmer C. and Bychawski Z., *Teoria Konstrukcji Sprezonych*, Panstwowe Wydawnictwo Naukowe, Warsaw, 1961 (in Polish).

Onoufriou T. and Frangopol D.M., Reliability-Based Inspection Optimization of Complex Structures: A Brief Retrospective, *Computers and Structures*, Vol. 80(12), 1133-1144, 2002.

Östberg G.H., On the Meaning of Probability in the Context of Probabilistic Safety Assessment, *Reliability Engineering and System Safety*, Vol. 23, 305-308, 1988.

Otstanov V.A. and Reizer V.D., Accounting for the Responsibility of Buildings and Structures in the Building Codes, *Stroitelnaya Mekhanika I Raschet Soorunzhenii*, No.2, 11-14, 1981 (in Russian).

Owen D.B., Craswell K.J. and Hanson D.L., Nonparametric Upper Confidence Bounds for $P(X<Y)$ and Confidence Limits for $P(X<Y)$ when X and Y are Normal, *Journal of the American Statistical Association*, Vol. 59, 906-924, 1964.

Paez A. and Torroja E., La Determinicion del Coeffocoente de Seguridad en las Distintas Obras, *Instituto Technico del la Construccion y del Cemento*, Madrid, 1959 (In Spanish).

Paloheimo H., Structural Design Based on Weighted Fractals, *Journal of the Structural Division*, Vol. 100,1367-1378, 1974.

Paloheimo E. and Mannus M., Structural Design Based on Weighted Fractals, *Journal of the Structural Division*, Vol.100, 1367-1378,1974.

Pandey M. and Upadhyay S.K., Reliability Estimation in Stress-Strength Models: A Bayes Approach, *IEEE Transactions on Reliability*, Vol. R-35, No. 1, p.98, 1986.

Park J.W. and Clark G.M., A Computational Algorithm for Reliability Bounds in Probabilistic Design, *IEEE Transactions on Reliability*, Vol. R-35, 3031, 1986.

Parkinson D.B., Computer Solution for the Reliability Index, *Engineering Structures*, Vol. 2, 57-62, 1980.

Parkinson D.B., Alternative Reliability Formats, *Engineering Structures*, Vol. 5, 207-214, 1983.

Paté-Cornell M.E., Uncertainties in Risk Analysis: Six Levels of Treatment, *Reliability Engineering and System Safety*, Vol. 54, 95-111, 1996.

Peck D.S., Uses of Semiconductor Life Distributions, in *Semiconductor Reliability*, Vol.2 pp.10-22, Engineering Publishers, Elizabeth, New Jersey, 1962.

Pell W.H., Limit Design of Plates-The Upper Bound for the Safety Factor, *Bulletin of American Mathematical Society*, Vol. 58(2), 195-199, 1952.

Pendola M., Hornet P., Mohamed A. and Lemaire M., Uncertainties Arising in the Assessment of Structural Reliability, *13th ASCE Engineering Mechanics Conference*, The Johns Hopkins University, Baltimore, MD, June 13-16, 1999.

Peruval N., Is Robot Technology Safe?, pp. 82-85, in *Decade of Robotics*, Springer-Verlag and IFS Publications Ltd., Bedford, UK, 1983.

Petroski H., *Design Paradigms*, Cambridge University Press, 1994.

Petroski H., *Invention by Design*, Harvard University Press, Cambridge, MA,1996.

Peters R.I., *Defining Design Safety*, Machine Design, p.128, Feb.1968.

Picon A., Navier and the Introduction of Suspension Bridges in France, *Construction History*, Vol. 4, 21-34, 1988.

Pikovsky, A. A., *Statics of Column Systems with Compressed Elements*, Gosudarstvennoe Izdatel'stvo Fiziko-Matematicheskoy Literatury, Moscow, 1961 (in Russian).

Pinto J.T., Blockley D.I. and Woodman N.Y., The Risk of Vulnerable Failure, Structural Safety, Vol.24, 107-122, 2002.

Platt J.T. and Wooller R., Sensitivity of Combat Aircraft Mass to Changes in Structural Design Safety Factures, SAWE Paper No. 1203, *37th Annual Conference of the Society of Allied Weight Engineers*, May 1978.

Pochtman Y.M. and Pystigsrodskii Z-I., Safety Factor of Certain Optimum Structures Made of an Ideal Elastoplastic Material, *Problemy Prochnosti*, No. 11,47, 1973.

Pomade P., Approach for Obtaining the Actual Safety Factor in an Elementary Model Consisting of a Cohesive Granular Material Subjected to Unidirectional Compression from the Average Stress between Two Singular Festures, *Materials and Structures,* Vol. 27(168), 196-205, 1994.

Popp H.G., Reliability through Statistical Material Property Definition, *SAE Paper 580 B*, Oct. 1962.

Pradlwarter H.J., Schuëller G.I, Jehlicka P. and Steinhiblertt, Structural Failure Probabilities of the HDR-Containment, *Journal of Nuclear Engineering and Design*, Vol. 128, 237-246, 1991.

Pronikov A. S., *Reliability of Machines*, Moscow, Mashino-strolnie Publishers,1978 (in Russian).

Prot M., Note sur le notion de coefficient de securite, *Annale des Ponts et Chaussees*, No. 27, 1936 (in French).

Prot M., La determiniation ratonelle et le controle des coefficients de securite, *Travaux*, No.222,1953 (in French).

Pugsley A.G., Concept of Safety in Structural Engineering, *Journal, Institute of Civil Engineers*, London, Vol. 36(5), 5-31, 1951.

Pugsley A.G., Structural Safety, *Journal Royal Aeron.Soc.*,Vol.59,1955.

Pugsley A.G., Current French in the Specifications of Structural Safety, *The Engineer*, 595-596, June 1956.

Pugsley A., The Prediction Pronness to Structural Accidents, *Structural Engineering*, Vol.51, 195-196,1973.

Punhes K. Airplane Reliability in a Nutshell, *IEEE Transactions on Reliability*, Vol, R-32, No. 2, 130-133, 1983.

Qu X. and Haftka R.T., Response Surface Approach Using Probabilistic Safety Factor for Reliability-Based Design Optimization, *Proceedings 2nd Annual probabilistic Method Conference*, Newport Beach, CA, 2002

Qu X. and Haftka R.T., Reliability-Based Design Optimization using Probabilistic Safety Factor, *Paper AIAA-2003-1657, 44 AIAA/ASME/ASCE/AHS/ASC Structures, Structural Dynamics, and Materials Conference*, Norfolk, VA, 2003

Quin S. and Widera G.E.O., Use of Stress-Strength Model in Determination of Safety Factor for Pressure Vessel Design, *Journal of Pressure Vessels Technology*, Vol. 118(1), 27-32, 1996.

Rabon L.M. Jr., Mechanical-System Reliability Testing, *Proceedings of Annual Reliability and Maintainability Symposium*, pp. 278-282, IEEE Press, New York, 1981.

Rackwitz R., Practical Probabilistic Approach to Design, Bull. 112, *Comité Européen du Béton*, France, 1976.

Rackwitz R., Structural Reliability Methods--Solutions and Problems in Probability Integration, in *Sixth International Conference on Applications of Statistics and Probability in Civil Engineering*, (Esteva L., and Ruiz S.E., eds.), Vol.3, pp.65-78, Mexico City, 1991.

Rackwitz R., Reliability Analysis--Past, Present and Future, *Proceedings of 8th ASCE Speciality Conference on Probabilistic Mechanics and Structural Reliability*, Paper No. PMC-200, 2000.

Raizer V.D., *Reliability Theory Methods for Design Code Making*, "Stroyzdat" Publishing, Moscow, 1986 (in Russian).

Raizer V.D., *Analysis of Structural Safety and Design Code Making Procedures*, "Stroyzdat" Publishing, Moscow, 1996 (in Russian).

240

Raizer V.D., Personal Communication, March 1, 2002.

Randall F.A., Composition of Factors of Safety, Annotated Special Bibliography, No. 194, Portland Cement Association, Skokie, Illinois, 1969.

Ramsberg J., Comments on Bohenen blust and slovic, and Vrijling, van Hengel and Houben, *Reliability Engineering and Safety Systems*, Vol.67, 205-209, 2000.

Randall F. A. Jr., Composition of Factors of Structural Safety, *Special Bibliography No.194*, Research and Development Division Library, Technical Services Department, Portland Cement Association, Skokie, Ill., July 1969.

Randall F.A., The Safety Factor of Structures in History, *Professional Safety 12-18*, Jan, 1976.

Randall F. A., Jr., Historical Notes on Structural Safety, *ACI Journal*, Vol.70,669-679, 1973.

Randall F. A. Jr., Private Communication, August 11, 2003.

Rao C.V.S.K., Safety of Glass Panels against Wind Loads, *Engineering Structures*, Vol. 6, 232-234, 1984.

Rao I.R., and Hsieh C.H., The Choice of Type of Extreme Value Distribution, in *Sixth International Conference on Applications of Statistics and Probability in Civil Engineering*, (Esteva L., and Ruiz S.E., eds.), Mexico City, pp.48-55, 1991.

Rao S. S., *Reliability – Based Design*, McGraw-Hill, New York, 1992.

Ratynski A., Selling Your Reliability Program to the Air Force Electronic Systems Division, *Proceedings of Eleventh National Symposium on Reliability and Quality Control*, pp. 542-556, IEEE Press, New York, 1965.

Ravindra M. K. and Lind N. C., Trends in Safety Factor Optimization, in *Beams and Beam Columns* (Narayanan R., ed.), Applied Science Publishers, Barking, pp. 207-236, 1983.

Ravindra M.K., Heaney A.C. and Lind N.C., Probabilistic Evaluation of Safety Factors, *Final Report*, Symposium on Concepts of Safety of Structures and Methods of Design, London 1969, International Association for Bridge and Structural Engineering, Zurich, Switzerland, pp. 35-46, 1969.

Ravindra M.K., Lind N.C. and Siu W., Illustration of Reliability-Based Design, *Journal of Structural Division*, Vol.100, 1974.

Ray J.N., Safety Factor Evaluation of Single Piles in Sand, *Journal of Geotechnical Engineering Division*, Vol.102, 1093-1108, 1976.

Redler W.M., Mechanical reliability Research in the NASA, *Proceedings of Fifth Reliability and Maintainability Conference (Annals of Reliability and Maintainability, Vol. 5)*, pp.763-768, AIAA Press, New York, 1966.

Redler W.M., Mechanical Reliability Research in the National Aeronautics and Space Administration, *Proceedings of Fifth Reliability and Mantainability Conference (Annals of Reliability and Maintainability, Vol.5)*, pp763-768, AIAA Press, New York.

Reese R.C., Probabilistic Approaches to Structural Safety, *Journal of the American Concrete Institute*, Vol. 79, 37-49, 1970.

Reethof G., State-of-the-Art: Mechanical and Structural Reliability, *Proceedings of Reliability and Maintainability Conference (Annals of Reliability and Maintainability, Vol. 10)*, p.62, ASME Press, New York, 1971.

Reiser B, A Remark on Ichikawa's Upper Bound of Probability Failure, *Reliability Engineering*, Vol. 13, 181-183, 1985.

Reiser B, There is No Such Thing as a Free Lunch: A Comment on a New Method for Reliability Demonstration, *Reliability Engineering*, Vol. 13, 175-180, 1985.

Reiser B. and Faraggi D., Confidence Bounds for Pr(a'x>b'y), *Statistics*, Vol. 25, 107-111, 1994.

Reiser B., Stress-Strength Models, *Encyclopedia of Statistical Science*, Vol. 9, 15-16, 1988.

Reiser B. and Faraggi D., Confidence Intervals for the Generalized ROC Criterion, *Biometrics*, Vol. 53, 644-652, 1997.

Reiser B. and Guttman I., Statistical Interference for P(Y<X): The Normal Case, *Technometrics*, Vol. 28, 253-258, 1986.

Reiser B. and Guttman I., A Comparison of Three Point Estimators for P(X<Y) in the Normal Case, *Computational Statistic and Data Analysis*, Vol. 5, 59-66, 1987.

Reiser B. and Guttman I., Sample Size Choice for reliability Verification in Strength Stress Models, *The Canadian Journal of Statistics*, Vol. 17, 253-259, 1989.

Reiser B., Guttman I. and Faraggi D., Choice of Simple Size for Testing the *P(X<Y)*, *Communications in Statistics. Theory and Methods*, Vol. 21(3), 559-569, 1992.

Reiser B. and Rocke D, Interference for Stress Strength Problems under the Gamma Distribution, *Journal of the Indian Association for Productivity*, Quality and Reliability, Vol. 18, 1-22, 1993.

Reitman M.I., *Zalog Prochnosti (Warrant for Strength)*, "Stroiizdat" Publishers, Moscow, 1979 (in Russian).

Reshetov D., Ivanov A., and Fadeev V., *Reliability of Machines*, Mir Publishers, Moscow, 1990.

Richard F. and Perreux D., The safety-Factor Calibration of Laminates for Long-Term Applications: Behavior Model and Reliability Method, *Composite Science Technology*, Vol. 61(14), 2087-2094, 2001.

Roark R.J., Just How Safe Are You?, *Product Engineering*, 97-101, Aug. 1965.

Rocha M., Serafin J. L. and Estevas- Ferreira M. J., The Determination of the Safety Factors of Arch Dams by Means of Models, *Bulletin Rilem No. 7*, Memoria No. 163.

Rockafeller R.T., and Wets W.J.-B., Scenarios and Policy Aggregation in Optimization under Uncertainty, *Mathematics of Operations Research*, Vol.16,119-147,1991.

Roenheim D.E., Analysis of Reliability Improvement through Redundancy, Proceedings of the New York University Conference on Reliability Theory, pp. 119-142, New York & University, 1958.

Roësset J. and Yao J.T.P., Civil Engineering Needs in the 21st Century, *Journal of Professional Issues Engineering*, Vol. 114/3, 248-255, 1988.

Roësset J. and Yao J.T.P., State of the Art of Structural Engineering, *Journal of Structural Engineering*, Vol. 128(8), 965-975, 2002.

Rosenblueth E., Design Philosophy: Structures, in *Proc. 2nd Int. Conf. Application of Statistics and Problems in Soil and Structural Engineering*, Aachen, West Germany,1975.

Rosenblueth E., What Should We Do with Structural Reliabilities?, in *Reliability and Risk Analysis* (Lind N.C., ed.), University of Waterloo Press, pp.24-34, 1987.

Rosenblueth E., Here and Henceforth, in *Sixth International Conference on Applications of Statistics and Probability in Civil Engineering*, (Esteva L., and Ruiz S.E., eds.), Vol.3,pp.81-94, Mexico City, 1991.

Rosenfeld M. J., Vieth P. H., and Haupt R. W., Proposed Corrosion Assessment Method and In-Service Safety Factors for Process and Power Piping Facilities, *ASME Pressure Vessels and Piping Division Publications*, Vol. 353, pp. 395-405, ASME Press, New York, 1997.

Ross Sh., *A First Course in Probability*, Prentice Hall, Upper Saddle River, NJ, p.396, 1998.

Row R.E., Current European Views on Structural Safety, *Journal of the Structural Division*, Vol. 96, 461-467, 1970.

Rowe R.E., Current European Views on Structural Safety, *Journal of Structural Division*, 461-467, 1970.

Rowe W. D., *The Anatomy of Risk*, Robert E. Krieger Publishing Company, Malabar, FL, 1983.

Rusch H., Der Einfluss des Sicherheitsbegriffes auf die technischen Regeln fur vorgespannten Beton, *Schweiz. Arch.*, No.1, 1954 (in German).

Rzhanitsyn A.R., Determination of Safety Factors in Construction, *Stroitel'naya Promishlennost*, No. 8, 1947 (In Russian).

Rzhanitsyn A. R., *Design of Construction with Materials' Plastic Properties Taken into Account*, Gosudarstvennoe Izdatel'stvo Literatury Po Stroitel'stvu i Arkhitekture, Moscow,1949 (First edition) 1954, (second edition), Chapter 14 (in Russian) (see also a French translation: A. R. Rjanitsyn: *Calcul à la rupture et plasticité des constructions*, Eyrolles, Paris, 1959).

Rzhanitsyn A.R., It is Necessary to Improve the Norms of Analysis of Building Constructions?, *Stroitelnaya Promyshlennost*, No.8, 1957 (in Russian).

Rzhanitsyn A.R., Determination of the Safety at Loads, Representing Random Processes, *Stroitelnaya Mekhanika i Raschet Sooruzhenii*, No. 3, 7-11, 1971, (in Russian).

Rzhanitsyn A.R., Economic Principle of Design for Safety, *Stroitelnaya Mekhanika i Raschet Sooruzhenii*, No.3, 3-5,1973 (in Russian).

Rzhanitsyn A. R., *Theory of Reliability Design of Civil Engineering Structures*, "Stryoizdat" Publishing House, Moscow, 1978 (in Russian).

Sacchi G. Upper and Lower Boends for the Average Value and for the Variance of the Safety Coefficients of Structures with Random Resistance, *Mechanics*, 101-103, June 1971.

Sachs H. M., Reasons, Results, and Remedies for Pump Safety Factors Overuse, *ASHRAE Transactions*, Vol.107, 559-565, 2001.

Sagot J.C., Gouin V. and Gomes S., Ergonomics in Product Design: Safety Factor, *Safety Science*, Vol. 41(2-3), 137-154, 2003.

Sanford A.G., Stress Test and Safety Factors in Structural Gazing, *Constructor Specifier*, Vol. 40, 31-32, 37 1987.

Sarton G., *A Guide to the History of Science* (Horus), Chronica Botanica Co., Waltham, Ma., 1952.

Sathe Y.S. and Shah S.P., On Estimating $P(X>Y)$ for the Exponential Distribution, *Communications in Statict. A: Theory and Methods*, Vol. 10, 339-347, 1981.

Savchuk V.P., On the Problem of Safety Factor Foundation, *Izvestiya Vuzov, Aviatsionnaya Tekhnika*, No.1, 29-32, 1989.

Savoia M., Structural Reliability Analysis through Fuzzy Number Approach, with Application to Stability, *Computers and Structures*, Vol. 80, 1087-1102, 2002.

Schatz R., Shooman M.L. and Shaw L., Application of Time Dependent Stress-Strength Models of Non-Electrical and Electrical Systems, *Proceedings of Annual Reliability and Maintainability Conference*, pp.598-604, IEEE Press, New York, 1966.

Schatz R., Scooman M. and Shaw L., Application of Time Dependent Stress-Strength Models of Non-Electrical and Electrical Systems, in *Proc. Annual Reliability and Maintainability*, Symp., pp. 540-547, 1974.

Scherer R.Y. and Schuëller G.I, On the Data Bank of the Friuli Earthquake, *European Earthquake Engineering Journal*, Vol.2, 437-451, 1988.

Schigley J. E., Discussion on the Paper by C. Mischke, *Journal of Engineering for Industry*, 541-542, Aug. 1970.

Schiller H. and Stalberg E., Safety Factor in Neuromuscular Transmission, *Schweizarische Arch. Neurologie*, Vol. 120(1), 54-64, 1977.

Schoof R.F., How Much Safety Factor?, *Allis- Chalmers Electrical Review*, 21-24, 1960.

Schorn G. and Lind N.C., Adaptive Control of Design Codes, *Journal of Engineering Mechanics*, Vol. 100, 1-16, 1974.

Schuëller, G. I., *Einführung in die Sicherheit und Zuverlässigkeit von Tragwerken*, Verlag von Wilhelm Ernst & Sohn, Berlin, 1981 (in German).

Schuëller G.I., A Prespective Study of Materials Based on Stockastic Methods, *Material and Structures – Matériaux et Contructions*, Vol. 20, 242-247, 1987.

Schuëller G.I., Recent Developments in Structural Computational Stohastic Mechanics, in *Computational Mechanics for the Twenty First Century* (Toppng B.H.V., ed.), pp.281-310, Saxe-Coburg Publishers, Edinburg, 2000.

Schuëller G.I. and Ang A.H-S., Advances in Structural Reliability, *Nuclear Engineering and Design*, Vol. 134, 121-140, 1992.

Schuëller G.I and Freudenthal A.M., Scatter Factor and Reliability of Structures, *NASA CR-2100*, 1972.

Schuëller G.I. and Kafka P., Future of Structural Reliability Methodology in Nuclear Power Plant Technology, *Nuclear Engineering and Design*, Vol. 50, 201-205, 1978.

Schuëller G.I. and Strix R., A Critical Appraisal of Methods to Determine Failure Probabilities, *Journal of Structural Safety*, Vol.4,293-309, 1987.

Schultze E., The General Significance of Statistics for the Civil Engineer, *in Proc. 2nd Int. Conf. Application of Statistics and Probability in Soil and Structural Engineering*, Aachen, West Germany, 1975.

Show F.S., The Safety Factor of an Elastic-Plastic Body in Plane Strain, *Journal of Applied Mechanics*, Vol. 19(2), 232-233, 1952.

Sedrakian L.G., *Towards Statistical Theory of Strength*, Ereven, 1958 (in Russian).

Sekhniashvili E. A., Napetvaridze Sh. G. and Losaberidze A.A., The Life and Activity of Academician G.M. Mukhadze, in *Selected Works* by Mukhadze G., Metsniereba Publishing House, pp. 7-27, Tbilisi, 1986 (in Georgian).

Sementsov S.A., About Possibility of Determining the Reliability of Building Construction by Probabilistic Methods, *Stroitelnaya Mekhanika i Raschet Sooruzhenii*, No.4.2-7,1972 (in Russian).

Sen P.K., A Note on Asymptotically Distribution-Free Confidence Bound for $P(X<Y)$ Based on Two Independent Samples, *Sankhyā Series A*, 29, Part 1, 95-102, 1967.

Seneta E., Bienayme, Irenne-Jules, in *Leading Personalities in Statistical Sciences* (Johnson N.L. and Kotz S., eds.), pp. 21-23, Wiley, New York, 1997.

Seneta E., Markov, Andrei Andreevich, in *Leading Personalities in Statistical Sciences* (Johnson N. and Kotz S., eds.), pp. 263-265, Wiley, New York, 1997.

Serensen S.V, Kogaev V.P. and Shneiderovich R.M., *Load Carrying Capacity and Strength Analysis of Machine Parts*, Second edition, "GNTIML" Publishers, Moscow, 1963 (in Russian).

Serensen S.V. and Kogaev V.P., Probabilistic Analysis on Strength, *Vestnik Mashinostroenia*, No.1, 1968 (in Russian).

Serensen S.V. and Kozlov L.A., Characteristics of Nonstationary Stress and Determination of a Safety factor, *Vestnik Mashinostroenia*, No.6, 1964 (in Russian).

Shah H.C. and Tang W.H., Statistical Evaluation of Load Factors in Structural Design in *Proceedings of Ninth Reliability and Maintainability, Annals of Reliability and Maintainability*, Vol. 9, pp. 650-658, SAE Press, Warrendale, PA, 1970.

Shaw L., Shooman M.L. and Schatz R., Time-Dependent Stress-Strength Models for Non-Electrical and Electrical Systems, *Proceedings of Annual Reliability and Maintainability Symposium*, pp.186-197, IEEE Press, New York, 1973.

Shen K., On the Relationship Between Component Failure Rate and Stress-Strength Distributional Characteristic, *Microelectronics and Reliability*, Vol. 28, 801-812, 1988.

Sherman N. (ed.), The Stone Edition, *The Torah*, Mesorah Pubblications, Ltd., New York, p.1051, 1993.

Shinozuka, M., Basic Analysis of Structural Safety, *Journal of Structural Engineering*, Vol. 59, No. 3, 721-740, 1983.

Shinozuka M., and Nishimura A., On General Representation of Density Function, *Annals of Reliability and Maintainability*, Vol. 4, 897-903, 1965.

Shinozuka M. and Yang J-N., Optimum Structural Design Based a Reliability and Proof-Load Test, in *Proceedings of Eighth Reliability and Maintainability Conference* (1969 Annals of Assurances Sciences), pp. 375-391, Gorden and Breach Science Publishers, New York, 1969.

Shiraishi N., Furuta M. and Sugimoto M., Effects of Group Errors in Structural Design, *Structural Safety*, Vol. 2, 245-250, 1985.

Shkinev A.N., Analysis of the Causes of Accidents as a Factor of Increasing Structural Reliability, *Stroitelnaya Mekhanika i Raschet Sooruzhenii*, No.1.13-15,1976 (in Russian).

Shnurer H., Future Developments of Probabilistic Structural Reliability to Meet the Needs of Risk Analysis of Nuclear Power Plants, *Nuclear Engineering and Design*, Vol. 60, 145-149, 1980.

Shoof R., How Much Safety Factor?, *Allis-Chalmers Electrical Review*, # 21-24, 1960.

Shooman M.L., *Probabilistic Reliability: An Engineering Approach*, McGraw-Hill, New York, 1968.

Shooman M.L., Reliability Phisics Models, *IEEE Trans. Reliab.*, Vol.R-17(1), 14-28, 1968.

Shrader-Rechette K.S., *Risk Analysis and Scientific Method*, D. Reidel, Holland, 1985.

Shukailo V.F., Classification of the Models "Loading-Strength" and Justification of the Types of Distribution Function of the Margins of Mechanical Elements, in *Reliability Problems in Structural Mechanics* (V.V. Bolotin and A.A. Chiras, eds.), pp. 86-92, "RINTIP" Publishers, Vilnius, 1968 (in Russian).

Saint-Venant A.B. de, C. Navier, Résumé des leçons donnée a l'Ecole des ponts et chaussées, Paris, 1864 (in French).

Silver E.A., Safety Factor Approximation Based on Turkeys Lambda Distribution, *Operations Research Quarterly*, Vol. 28(3), 743-746, 1977.

Simonoff J., Hochberg Y. and Reiser B., Alternative Estimation Procedures for $P(X<Y)$ in Categorized Data, Biometrics, Vol. 42, 895-907, 1986.

Singer J., Abramovich H., and Yaffe R., Initial Imperfection Measurements of Integrally Stiffened Cylindrical Shells, *TAE Report 330*, Department of Aeronautical Engineering, Technion-Israel Institute of Technology, Haifa, Israel, 1978.

Singh N., MVUE of $P(X<Y)$ for Multivariate Normal Populations: An Application to Stress-Strength Models, *IEEE Transactions on Reliability*, Vol. R-32, No. 2, 214-216, 1983.

Smith C.O., *Introduction to Reliability in Design*, McGraw-Hill, New York, 1976.

Smith C.O., Design of Ellipsoidal and Toroidal Pressure Vessels to Probabilistic Criteria, *Journal of Mechanical Design*, Vol. 102, 787-792, 1980.

Smith G. N., *Probability and Statistics in Civil Engineering*, Collins Professional and Technical Books, London, 1986.

Smith I.F.C. and Hirt M.A., Fatigue Reliability: ECCS Factors, *Journal of Structural Engineering*, Vol. 113, 623-628, 1987.

Smith R. L., Estimating Tails of Probability Distribution, *Amn. Statistics*, Vol.15, 1174-1207, 1987.

Solomon K.I., Chazen C. and Miller R.L., Compilation and Review - The Safety Factor, *Journal of Accountancy*, Vol. 156(1), p. 50, 1983.

Soloviev A.P., About the Safety Factor of a Structure, Designed via the Prototype, *Stroitelnya Mekhanika i Raschet Sooruzhenii*, No.5, 1969 (in Russian).

Stancampiano P. A., Monte Carlo Approaches to Stress-Strength Interference, in *Failure Prevention and Reliability* (Bennett S. B., Ross A. L. and Zemanick P. Z., eds.), ASME Press, New York, pp. 197-212, 1977.

Stancampiano P.A., Monte Carlo Approaches to Stress Strength Interference, *Failure Prevention and Reliability*, pp.285-309, ASME Press, New York, 1977.

Stancampiano P.A., Some Thoughts on the Future of Probabilistic Structural Design of Nuclear Components, *Nuclear Engineering and Design*, Vol. 50, 207-211, 1978.

Stanley F.R., Historical Note on the 1.5 Factor of Safety for Aircraft Structures, *Journal of Aerospace Science*, Vol. 29(2), 243-244, 1972.

Starr C. et al, Philosophical Basis for Risk Analysis, *Annual Review of Energy*, Vol.1, 629-662, 1976.

Starr C. and Whipple Ch., Risk and Risk Decisions, *Science*, Vol. 208, 1114-1119, 1980.

Steinberg A., NASA Program Decisions using Reliability Analysis, *Proceedings, 1972 Annual Reliability and Maintainability Symposium*, 458-473, 1972.

Stevens M.A., Simons D.B. and Lewis G. L., Safety Factors for Riprap Protection, *Journal of Hydraulic Division*, Vol. 102, 637-655, 1976.

Straub H., *A History of Civil Engineering*, Leonard Hill Limited, London, 1952.

Streletskii N.S., Towards the Analysis of theGeneral Safety Factor, *Proekt i Standart*, No.10, 1935 (in Russian).

Streletskii N.S., The Factor of Safety as an Indicator of Strength Equality of Structures, *Comptes Rendus Doklady de L'Academie des Sciences de L'URSS*, Vol. XIV, No.8, 487-489, 1937.

Streletskii N.S., About a Possibility of Increasing the Allowable Stresses, *Stroitelnaya Promyshlennost*, No.2-3, 1940 (in Russian).

Streletskii N.S., About a Question of Determining the Allowable Stresses, *Stroitelnaya Promyshlennost*, No.7, 1940 (in Russian).

Streletskii N.S., Practical Consequences of Statistical Analysis of Safety Factor, *Stroitelnaya Promishlenmost*, No.10, 5-8, 1946 (in Russian).

Streletskii N.S., About Establishment of Safety Factors of Structures, *Izvestiya Akademii Nauk SSSR, Otdelenie Tekhnicheskikh Nauk*, No.1,15-26, 1947 (in Russian).

Streletskii N. S., *Statistical Basis of the Safety Factor of Structures*, "Stroyizdat" Publishing House, Moscow, 1947 (in Russian).

Streletskii N.S., Modern State of the Design Problem of Construction, *Izvestiya Visshikh Uchebnykh Zavedenii, Stroitelstvo I Arkhitektura*, No.1, 3-11, 1963 (in Russian).

Streleskii N.S., *Izbrannye Trudy* (Selected Works) (Belenia E.I., ed.), "Stroiizdat" Publishers, Moscow, 1975 (in Russian).

Stüssi F., *Schweizarische Bauzeitung*, Vol. 116, p. 201, 1940 (in German).

Su H. L., Statistical Approach to Structural Design, *Proceedings of the Institution of Civil Engineers*, Vol. 13, 353-362, 1959.

Su H.L.,Philosophical Aspects of Structural Design, *Journal of Structural Division*, Vol.87 (5), 1961.

Subramanian R. and Ramesh K.S., A Note on Stress-Strength Reliability Models, *IEEE Transactions on Reliability*, Vol. R-18, No. 4, 204-205, 1969.

Sudakov R.S. and Chekanov A.N., Towards Determination of Safety Factor of Unique Mechanical Structures under Limited Information, *Izvestiya Vyschikh Uchebrikh Zavedenil, Mashinostroenie*, No.5, 1971 (in Russian).

Sudakov R.S. and Chekanov A.N., Determination of Structural Reliability via Probabilistic Models of the "Load-Strength" Type, in *Reliability and Quality Control*, pp.47-55, Standards Publishing, Moscow, 1972 (in Russian).

Suensson N. L., Factor of Safety Based on Probability, *Engineering*, 414-155, Jan. 1961. *Symposium on Reliability*, pp. 240-247, IEEE Press, New York, 1965.

Sundararajan C., Probabilistic Assessment of Pressure Vessel and Piping Reliability, *Journal of Pressure Vessel Technology*, Vol.108, 1-13, 1986.

Sundararajan C., *Probabilistic Structural Mechanics Handbook*, Chapman and Hall, New York, 1995.

Sundararajan C. and Witt F.J., Stress-Strenght Interference Method, in *Probabilistic Structural Mechanics Handbook* (Sundararajan C., ed.), pp.8-26, Chapman and Hall, New York, 1995.

248

Suppes P., and Zanotti M., On using Random Relations to Generate Upper and Lower Probabilities, *Synthese*, Vol.36, 427-440, 1977.

Szabo I., *Geschichte der Mechanischen Prinzipien*, Birkhäuser, Basel, 1983 (in German).

Szerszen M. and Nowak A.S., Calibration of Design Code for Buildings (ACI 318): Part 2 – Reliability Analysis and Resistance Factors, *ACI Structural Journal*, Vol. 100, 383-389, 2003.

Tal K.E., On Improvement of Principles of Reliability Determination of Building Structures, *Stroitelnaya Mekhanika i Raschet Sooruzhenii*, 1970 (in Russian).

Tal K.E., About Improving Principles of Determination of Structural Reliability, *Stroitelnaya Mekhanika i Raschet Sooruzhenii*, No.4.50-51,1975 (in Russian).

Taub J., Structural Safety of Aircrafts, in *Contributions to the Theory of Aircraft Structures*, Delft University Press, pp. 39-50, 1972.

Taylor A.E., Martin D. and Parker J.C., Tissue Resistance Safety Factor, *International Journal of Microcirculation*, Vol. 1(3), 223-233, 1982.

Tejchman A. and Gwizdals K., Analyse Des Coefficients De Securite de Capacite Pertante Des Pieux De Grand Diametre, *Proceedings of the Seventh European Conference on Soil Mechanics*, 293-296, 1979 (in French)

Thangjitham S. and Heller R.A., Reliability of Proof Loaded Fiber-Composite Plate under Randomly Oriented Loads, *AIAA/ASME/ASCE/AHS 28th Structures Structural Dynamics and Materials Conference*, pp. 275-281, Monterey, CA., 1987.

Thiruvengalsm A., Now There Is a Way to Work out Erosion Strength of Materials, *Product Engineering*, Vol. 37, p. 55, 1966.

Thoft-Christensen P., Fundamentals of Structural Reliability, in *Lectures on Structural Reliability* (Thoft-Christensen P., ed.), Institute of Building Technology and Structural Engineering, Aalborg University, Denmark, pp. 1-28, 1980.

Thoft-Christensen, P. and Baker, M. J., *Structural Reliability Theory and its Applications*, Springer Verlag, Berlin, 1982.

Thoft-Christensen P. and Sorensen J.D., Reliability of Structural Systems with Correlated Elements, *Applied Mathematical Modelling*, vol. 6, 171-178, 1982.

Thomas T.M., Derbalian G. and Bischel K., Parametric Sensitivity of Rocket Motor Reliability, *Proceedings for Annual Reliability and Maintainability Symposium*, pp. 121-128, IEEE Press, New York, 1980.

Tichý M. The Nature of Reliability Requirements, in *Methods in Stochastic Structural Mechanics* (Casciati F., and Faravelli L., eds.), pp. 153-165, SEAG, Pavia, 1985.

Tichý M., *Applied Methods of Structural Engineering*, Kluwer Academic publishers, pp.290, Dertrecht, 1993.

Tichý M., Engineering, Operational Economic and Legal Aspects of the Reliability Assurance, in *Reliability Problems* (Casciati F. and Roberts, J.B, eds.), pp. 137-160, Springer, Vienna, 1991

Tichý M., First-Order Third-Moment Reliability Method, *Structural Safety*, Vol.16, 189-200, 1994.

Tichý M., Foreword, in "Safety in Construction Works and Its Design According to Limit States Instead of Permissible Stresses" by M. Mayer, pp. 13-21, INTEMAC, Madrid, Spain, 1975.

Tichý M., Max Mayer – Begründer der Berechnungsmethode nach Grenzzuständen, Submitted on June 18, 1990 for publication in "Technische Mechanik", Magdeburg (in German).

Tichý M., Private Communication, May 16, 2003.

Tichý M., Private Communications, June 2003.

Tichý M. and Vorlicek, M., *Statistical Theory of Concrete Structures*, "Academia" Publishing House, Prague, 1972.

Tien M.D. and Massoud M., On the Relation Between the Factor of Safety and Reliability, *Journal of Engineering Industry*, Vol. 89(3), 853-857, 1974.

Tiffany C.F., On the Prevention of Delayed Time Failures of Aerospace Pressure Vessels, *Journal of the Franklin Institute*, Vol. 290, 567-582, 1970.

Tiger B. and Weir K., Stress-Strength Theory and Its Transformation into Reliability Functions, *Physics of Failure in Electronics*, Vol. 2, pp.94-101, 1964.

Timashev S. A., *Reliability of Large Mechanical Systems*, "Nauka" Publishing House, Moscow, 1982 (in Russian).

Timmons A.R. and Henser R.E., A Study of First Day Space Malfunctions, *Proceedings of Tenth Reliability and Maintainability Conference (Annals of Reliability and Maintainability Symposium Vol. 10)*, pp. 90-99, ASME Press, New York, 1971.

Timoshenko S.P., *History of Strength of Strength of Materials*, McGraw-Hill, New York, 1952.

Timoshenko S. P. and Gere J. M., *Theory of Elastic Stability*, McGraw Hill, Auckland, 1963.

Todhunter L. and Pearson K., *A History of Elasticity and of the Strength of Materials*, Dover Publications, New York, 1960. (original publication, Cambridge University Press, 1886).

Tong H., A Note on Estimation of $P(Y<X)$ in the Exponential Case, *Technometrics*, Vol. 16, 625, 1975 (Errata, Vol. 17, 395, 1975).

Tormilill T.A., Methods for Calculating the Reliability Function for Systems Subjected to Random Stresses, *IEEE Transactions on Reliability*, Vol. R-23, No.4, 256-262, 1974.

Torroja E., Report on Super Imposed Loads and Safety Factors, *International Council for Building Research Studies and Documentation*, Paris, France, 1957.

Torroja E.O.,About Loadings and Safety Factors, in *Materials of International Conferences on Analyisi of Building Constructions*, "Gosstroyizdat" Publischer, Moscow, 1961 (in Russian).

250

Townsend A.R., Structural and Mechanical Reliability Technology Introduction, *Proceedings of Eighth Reliability and Maintainability Conference* (1969 Annals of Assurance Science), pp. 341-342, Gordon and Breach Science Publishers, New York, 1969.

Tseitlin V.I., Fedor Chenks D.G., Safety Factor Evaluation in Multi component Loading with Allowance for Scatter of Material Properties, *Problemy Prochnosti*, 31-33, Sept. 1979.

Turkstra C.J. and Daly M.J., Two Moment Structural Safety Analysis, *Canadian Journal of Civil Engineering*, Vol. 5, 414-426, 1978.

Tye, W., Factors of Safety- or a Habit?, *Journal of Royal Aeronautical Society*, Vol. 48, 487-494, 1944.

Ugata T. and Moriyama K., Simple Method of Evaluating the Failure Probability of a Structure Considering the Skewness of Distribution, *Nuclear Engineering and Design*, Vol.160, 307-319, 1996.

Undusk V., Safety Factor of Pillars, *Oil Shale*, Vol. 15(2), 157-164, 1998.

Ury H.K., On Distribution-Free Confidence Bounds for $\Pr\{Y<X\}$, *Technometrics*, Vol. 14, 577-581, 1972.

Vable M., *Mechanics of Materials*, Oxford University Press, New York, pp.129-130, 2002.

Val D.V.and Stewart M.G., Partial Safety Factors for Assessment of Existing Bridges, in *Applications of Statistics and Reliability* (Melchers, R.E. and Stewart M.G.,eds),pp.659-666, Balkema, Rotterdam,2000.

Van der Neut A, Some Remarks on the Fundamentals of Structural Safety, *AGARD – Rep. 155*,1957.

Veneziano D., New Index of Reliability, *Journal of Engineering Mechanics*, Vol. 105, 277-296, 1977.

Verderaine V., Aerostructural Safety Factor Criteria using Deterministic Reliability, *Journal of Spacecraft and Rockets*, Vol. 30 (2), 244-247, 1993.

Vidal C., De Mollerat T. and Klein M, Evaluation of Tests and Design Factors, in *Procedings, International Conference on Spacecraft Structures and Mechanical Testing*, Noordwijk, The Netherlands, 19-21 October, 1988 (also ESA SP-289, 1989).

Vinogradov O., *Introduction to Mechanical Reliability*, Hemisphere Publishing Corp., New York, p.54, 1991.

Virabov R.V. et al, Calculation of Strip Feed Step Error and Selection of Adhesion Safety Factor in Roller Mechanisms, *Soviet Engineering Research*, Vol. 3, 16-17, 1983.

Vrijling J.K., Some Considerations on the Acceptable Probability of Failure, *Proceedings, 5th International Conference on Structural Safety and Reliability* (Ang

A.H.-S, Shinozuka M. and Schuëller G.I., eds.), pp.1919-1926, ASCE Press, New York, 1989.

Vrijling J.K., Van Hengel W. and Houben R.J., Acceptable Risk as a Basis for Design, *Reliability Engineering and System Safety*, Vol. 59, 141-150, 1998.

Vrouwenvelder T., Stochastic Modelling of Extreme Action Events in Structural Engineering, *Probabilistic Engineering Mechanics*, Vol. 15, 109-117, 2000

Vizir P. L., Reliability of an Element of the System, in *Loads and Reliability of Civil Constructions*, "ZNIISK" Publishers, Moscow, 1973, pp. 26-42 (in Russian).

Vogt S., Possibilities of Checking the Reliability Assumptions Made during the Design of a Weapon System using the Results of Flight Tests, in *Operations Research and Reliability* (Grouchko D., el.), pp. 251-268, Gordon and Beach Science Publishers, New York, 1971.

Voigt W., *Ann. Phys.*, Vol 4, 567-591, 1901 (in German).

Volkov D. P., and Nikolayev S. N.*, Reliability of Construction Machines and Equipment*, Moscow, 1974 (in Russian).

Von Alven W.H.(ed.), *Semiconductor Reliability*, Engineering Publishers, Elizabeth, New Jersey, 1962.

Walker E.L., Uses and Abuses of Transit System Failure Rates, *Proceedings of the Annual Reliability and Maintainability Symposium*, pp. 294-299, IEEE Press, New York, 1978.

Wang L. and Grandhi R. V., Safety Index Calculations Using Intervening Variables for Structural Reliability Analysis, *Computers and Structures*, Vol. 59, pp. 1139-1148, 1996.

Weibull W., A Statistical Theory of the Strength of Materials, Proceedings, *Royal Swedish Institute of Engineering Research*, No. 151, Stockholm, Sweden, 1939.

Weibull W., A Statistical Distribution Function of Wide Applicability, *Journal of Applied Mechanics*, Vol. 18, pp. 293-297, 1951.

Weibull W., A Survey of " Statistical Effects " in the Field of Mechanical Failure, *Applied Mechanics Reviews*, Vol 5 (11), 449-451, 1952.

Whittaker I.C. and Gerharz J.J., A Feasibility Study for Verification of Fatigue Reliability Analysis, *The Boeing Company AFML-TR-70-157*, Sept. 1970.

Wierzbilcki W., Probabilistic and Semi-Probabilistic Method for the Investigation of Structural Safety, *Archiwum Mechaniki Stosowanej*, Vol.9 (6),1957.

Williams B.E., The Probability of Failure for Piping Systems, in *Failure Prevention and Reliability* (Loo F.T.C., ed.), ASME Press, pp.147-150, New York, 1981.

Wilson Ch. E., *Computer Integrated Machine Design*, Prentice Hall, Upper Saddle River, NJ, p. 22, 1997.

Wissel J.W., Human Error Potential in Today's Ballistic Missile Systems, *Proceedings of IAS Aerospace Systems Reliability Symposium*, pp. 213-214, 1962.

Wong F.S. and Yao J.T.P., Health Monitoring and Structural Reliability as a Value Chain, *Computer-Aided Civil Infrastructure Engineering*, Vol. 16, 71-78, 2001.

Wong W., *How did This Happen? Engineering Safety and Reliability*, Institution of Mechanical Engineers, London, 2002.

Wu Y-T, Shin Y., Sues R. and Cesare M., Safety-Factor Based Approach for probability-Based Design Optimization, AIAA Paper 2001-1522, 42nd AIAA/ASME/ASC/AHS/ASC Structures, Structural Dynamics and Materials Conference and Exibit, Seattle Washinton, 16-19 April 2001.

Wu Y-T., Efficient Methods for Mechanical and Structural Reliability Analysis and Design, *Ph.D. Dissertation*, The University of Arizona, 1984.

Wu Y-T. and Wirsching P.H., A New Algorithm for Structural Reliability Estimation, *Journal of Engineering Mechanics*, 1985.

Wu Y.-T.and Wirsching P.H., A New Algorithm for Structural Reliability Eotimstion, *Journal of Engineering Mechanics*,1985.

Wu Y-T. and Wang W., Efficient Probabilistic Design by Converting Reliability Constraints to Approximately Equivalent Deterministic Constraints, *Journal of Integrated Design and Process Sciences*, Vol. 2(4), 13-21, 1998.

Xie M. and Chen K., Some New Aspects of Stress-Strength Modelling, *Reliability Engineering*, Vol.33, 131-140, 1991.

Yadav R.P.S., A Reliability Model for Stress Versus Strength Problem, *Microelectronics and Reliability*, Vol. 12, 119-123, 1973.

Yadav R.P.S., Component Reliability under Environmental Stress, *Microelectronics and Reliability*, Vol. 13, 473-475, 1974.

Yamaguchi A., Kondo S. and Togo Y., A Case Study on Effectiveness of Structural Reliability Analysis in Nuclear Reactor Safety Assessment, *Reliability Engineering*, Vol. 5, 21-36, 1983.

Yang J.N., Application of Reliability Methods of Fatigue, Quality Assurance and Maintenance, Structural Safety and Reliability, (Schuëller G.I., Shinozuka M. and Yao J.T.P., eds.), pp. 3-18, Balkema, Rotterdam, 1994.

Yang J.N., Reliability Analysis of Structures under Periodic Proof Test in Service, *AIAA Journal*, Vol.14 (9), 1225-1234, 1976.

Yang J.N. et al, Reliability Analysis and Quality Assurance of Rocket Motor Case Considering Proof Testing, Paper AIAA-93-1392-CP, La Jolla, CA, 1993.

Yang J.S. and Nikolaidis E., Design of Aircraft Wings Subjected to Gust Loads: A Safety Index Based Approach, *AIAA Journal*, Vol. 29(5), 804-812, 1997.

Yang L. et al, Safety Factor and Reliability for Composite Laminates, *Acta Astronautica and Astronautica Sinica*, Vol.12, 631-634, 1991 (in Chinese).

Yao J.T.P., A General Note on Structural Reliability, *Nuclear Engineering and Design*, Vol. 50, 145-147, 1978.

Yao J. T. P., *Safety and Reliability of Existing Structures*, Pitman, Boston, 1985.

Yao J.T.P., Reliability Evaluation of Complex System, in *Risk Analysis* (Nowak A.S., ed.), pp.249-252, University of Michigan, Ann Arbor, 1994.

Yao J.M.T., Private Communication, July 18, 2003.

Yao J.T.P. and Yeh H.Y., Safety Analysis of Statistically Indeterminate Trusses, *Proceedings of Sixth Reliability and Maintainability Conference (Annals of Reliability and Maintainability, Vol. 6)*, pp. 54-62, SAE Press, Warrendale, PA, 1967.

Yarushkin B.F. and Shefel V.V., Evaluation of the Hot Cranking Resistance Safety Factor Well Metal in Multi-Pass Narrow Gap Welding, *Welding Production*, 24-26, Aug. 1984.

Yen B. C. and Tany W.H., Risk- Safety Factor Reaction for Storm Sewer Design, *Journal of Environmental Engineering Division*, Vol. 102, 509-516, 1976.

Young D. M., Stresses in Eccentrecally Loaded Steel Columns, *Publication of the International Association of Bridge Structural Engineers*, Vol. 1, 1932.

Zang T., Henush M., Hillburger M., Kenny S., Luckring J., Padula S. and Stroud J. W., Needs and Opportunities for Uncertainty-Based Multi-disciplinary Design Methods for Aerospace Vehicles, *NASA LaRC TM Report 1777*, 2001.

Zech B. and Wittermann F.H., Probabilistic Approach to Describe the Behavior of Materials – Part II, *Nuclear Engineering and Design*, Vol. 48, 575-584, 1978.

Zemanick P.P. and Witt F.J., An Engineering Assessment of Probabilistic Reliability Analysis, *Nuclear Engineering and Design*, Vol. 50, 173-183, 1978.

Zhu T.-L., A Reliability Based Safety Factor for Aircraft Composite Structures, *Computers and Structures*, Vol.48, 745-748, 1993.

Zimmermann P., Private Communications, July 25, 2003.

Zimmermann P., Private Communications, July 28, 2003.

Znamenskii E.M. and Sukhov Yu.D., About the Analysis of Structures with Given Level of Reliability, *Stroitelnaya Mekhanika i Raschet Sooruzhenii*, No. 2, 7-9, 1987 (in Russian).

Zou T., Mahadevan S., Mourelatos Z. and Meernik P., Reliability Analysis of Automotive Body-Door System, *Reliability Engineering and System Safety*, Vol. 78, 315-324, 2002.

Additional Bibliography

Alper M.E., Weight vs. Reliability – A Design Choice, California Institute of Technology, Jet Propulsion Lab., *Report 7PL TR32-110*, August 18, 1961.

Bouton I., Fundamental Aspects of Structural Reliability, *Aerospace Engineering*, Vol. 21(6), 66-67, 82, 84-86, 88, 90, 92-93, June 1962.

Bouton I. and Graff K.G., Development of Structural Design Criteria from Statistical Flight Data, Wright Air Development Division, *Report TR60-497*, March 1961.

Bouton I. and Trent D.J., The Unreliability of Structural Reliability, *American Society for Metals, Materials Engineering Congress and Exhibition*, Philadelphia PA, 1969 (paper No. P9-12.3).

Bouton I., Trent D.J. and Chenewth H.B., Design Factors for Structural Reliability, *Annals of Reliability and Maintainability*, Vol. 5, 229-235, AIAA Press, 1966.

Epstein A., Another Historical Note on the 1.5 Factor of Safety, *Journal of Aerospace Sciences*, Vol. 29, 759-769, 1962.

Freudenthal A.M., Safety and Safety Factors for Airframes, *Report 153*, NATO, Palais de Chaillot, Paris, France, 1957.

Freudenthal A.M., The Safety of Aircraft Structures, Wright Air Development Center, *Technical Report No. 57-131*, 1957.

Freudenthal A.M. and Payne A.O., The Structural Reliability of Airframes, Air Force Materials Laboratory, *Report AFML-TR-64-401*, 1965.

Freudenthal A.M. and Shinozuka M., Structural Safety Under Conditions of Ultimate Load Failure and Fatigue, Wright Air Development Division, *Report TR61-177*, October 1961.

Freudenthal A.M. and Wang P.Y., Ultimate Strength Analysis of Aircraft Structures, *Journal of Aircraft*, Vol. 7(3), 205-210, 1970.

Freudenthal A.M. and Yang P.Y., Ultimate Strength Analysis, Air Force Materials Laboratory, *Report AFML-TR-69-60*, March 1969.

Goldman G.M., Discussion on Safety Factor Requirements for Supersonic Aircraft Structures, *Transactions, ASME*, Vol. 79, 986-989, 1957.

Heath W.G., Factors of Safety – Should They Be Reduced?, In *Factors of Safety: Historical Development, State of the Art and Future Outlook*, NATO Advisory Group Aerospace Research and Development, *AGARD Report 661*, November 1977.

Heath W.G., Investigation of Safe-Life Fail-Safe Criteria for Space Shuttle, *NASA CR-112049*, May 1972.

Hoff N.J., Philosophy of Safety in the Supersonic Age, *AGARD Report*, 87, August 1956.

Lundberg B.K.O., Speed and Safety in Civil Aviation, International Council of the Aeronautical Sciences, 3rd *Congress Proceedings*, Stockholm, 1962.

Lundberg B.K.O., Aviation Safety and the SST, *Astronautics and Aeronautics*, Vol. 3(1), 22-29, 1965.

Lundberg B.K.O., How Safe Is Fail-Safe?, *Shell Aviation News*, No. 270, 14-19, 1960.

Mangurian G.N., The Aircraft Structural Factor of Safety, *Report 154*, NATO, Palais de Chaillot, Paris, France, 1957.

Mangurian G.N., Is the Present Aircraft Structural Factor of Safety Realistic?, *Aeronautical Engineering Review*, Vol. 13(9), 63-75, 1954.

Muller G.E. and Schmid C.L., Factor of Safety – USAF Design Practice, In *Factors of Safety: Historical Development, State of the Art and Future Outlook*, NATO Advisory Group for Aerospace Research and Development, *AGARD Report 661*, November 1977.

Pugsley A.G., A Philosophy of Airplane Strength Factors, *British Aeronaut. Res. Committee*, Reports and Memoranda, No. 1906, 1942.

Savchuk V.P., On the Safety Factor Justification Problem, *Soviet Aeronautics*, Vol. 32(1), 35-38, 1986.

Shanley F.R., Historical Note on the 1.5 Factor of Safety for Aircraft Structures, *Journal of Aerospace Sciences*, Vol. 29, 243-244, 1962.

Williams J.K., Safety Factors, *Journal of the Royal Aeronautical Society*, Vol. 60, 311, 1956.

Appendix A
Accuracy of the Hasofer-Lind Method

"Engineers and researchers active in the field of structural reliability tend to expend a disproportional amount of resources to deal with mathematical questions involving [reliability index] β, although their motivation is quite understandable. Indeed, finding the β value becomes the sole purpose of reliability analysis in many instances without carefully examining not only the assumptions for structural behavior and loading models, lent also the background against which the analysis is to be used."

M. Shinozuka (1989).

"Engineers usually like to manipulate numbers."

J.T.P. Yao (1985)

A.1 Introductory Comments

As is known, only a small fraction of the problem of the probabilistic theory of structures is capable of exact solution, while in most cases one must make do with approximate evaluation of the reliability (or, alternatively, of its complement to unity - the probability of failure) of the structure. For many years the most widely used technique was the so-called mean-value first-order second-moment method (Rzhanitsyn, 1954; Cornell, 1969) which consists in expanding the performance function $g(X_1, X_2, ..., X_n)$ in a Taylor series for the mean vector $\overline{X}(\overline{X}_1, \overline{X}_2, ..., \overline{X}_n)$ with the first term retained for evaluation of the mean value $\overline{g}$ (overbar signifying the mathematical expectation) and two terms for the calculation of the mean square deviation, σ_g. The performance function is constructed so that $g(X_1, X_2, ..., X_n) \leq 0$ implies failure and $g(X_1, X_2, ..., X_n) > 0$ implies a safe state of the structure, where the uncorrelated normal variables $X_1, X_2, ..., X_n$ describe the probabilistic properties of the structure and its environment. The safety index β is then chosen as $\beta = \overline{g}/\sigma_g$ with the attendant probability of failure

$$P_f = \Phi(-\beta) \tag{A. 1}$$

where $\Phi(\cdot)$ is the standard normal distribution function. This formula is exact, if $X_1, X_2, ..., X_n$ are independent normal design variables and the performance function is linear:

$$g(X_1, X_2, ..., X_n) = c_o + c_1 X_1 + c_2 X_2 + ... + c_n X_n \tag{A. 2}$$

the c_j's being deterministic parameters. If in addition X_j's are standard normal variables, i.e. with zero mean and unit variances, β represents the distance from the coordinate origin to the failure surface (see Fig. A1 for the two-dimensional case). Utilizing this result, Hasofer and Lind (1974) proposed to approximate the failure probability by transforming, the space of basic normal variables X and a corresponding space of

256

standard uncorrelated variables Z, linearizing the failure surface at the point, nearest the origin, and determining the failure probability as per Eq. (A.1.).

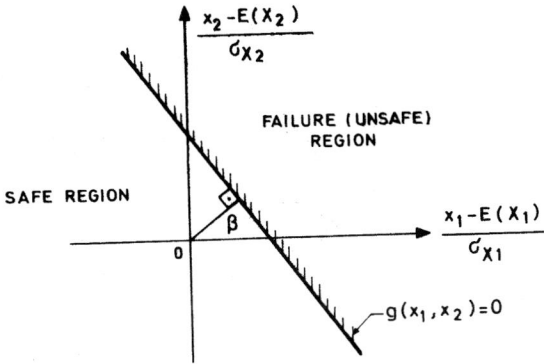

Fig. A. 1 Safe and unsafe regions: linear failure boundary

Thus,

$$\beta = \min_{Z \in \partial \omega} \left(\sum_{i=1}^{n} z_i^2 \right)^{1/2} \tag{A. 3}$$

where $\partial \omega$ is the failure surface in the z-coordinate system (Fig. A.2). It can be shown that the probability density of the standard uncorrelated variables is proportional to $e^{-r^2/2}$ where r is the distance from the origin.

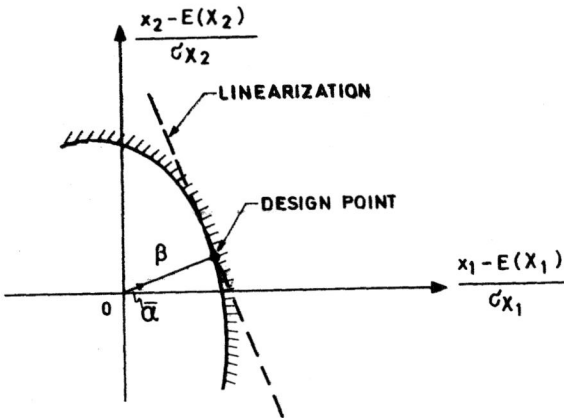

Fig. A. 2 Location of the design point for the nonlinear failure boundary

The linearization point thus has the maximum density among all points in the unsafe region, and the failure surface is accordingly best approximated in a neighborhood which makes a dominant contribution to the failure probability; differences between the true (nonlinear) and approximate (linear) surfaces in the neighborhoods far removed from the linearization point are immaterial, in view of the low densities involved.

The minimum distance in the general case of a nonlinear failure surface is found by an iterative method. The distance β and the unit vector $\bar{\alpha} = (\alpha_1, \ldots, \alpha_n)$ given by $O\vec{A} = \beta\bar{\alpha}$ where A is the design point (Fig. A.2.) can be determined by solving the following $n+1$ equations (Thoft-Christensen and Baker, 1982)

$$\alpha_i = \frac{-\left.\frac{\partial g}{\partial z_i}\right|_{z_i = \beta\alpha_i}}{\left[\sum_{k=1}^{n}\left(\left.\frac{\partial g}{\partial z_k}\right|_{z = \beta\alpha_k}\right)^2\right]^{1/2}}$$

(A. 4)

$$g(\beta\alpha_1, \beta\alpha_2, \ldots, \beta\alpha_n) = 0$$

While the Hasofer-Lind method has found widespread use in literature, to the best of our knowledge little attention has been paid so far to direct demonstration its accuracy. The latter is best checked with the aid of problems capable of exact solution. Shinozuka (1983) compared, in his extremely thorough paper an exact and an approximate solution of a sample problem first studied by Ellingwood (1980). The performance function was taken as $g(X_1, X_2, \ldots, X_n) = X_1 X_2 - X_3$, with X_1 distributed lognormally, X_2 normally and X_3 - causally. Shinozuka (1983) found excellent agreement between the exact solution and the first-order second moment method. Inter alia, he demonstrated that the Lagrange multiplier formalism, with the attendant numerical algorithms, can be successfully used to evaluate the safety index; moreover, it was established that the design point is the point of maximum likelihood. A closely related example with a performance function $g(X_1, X_2) = X_1^2 / X_2 - 1$, X_1 and X_2 being independent log-normal variates, was worked out recently by Ang and Tang (1984). Breitung (1983) derived an asymptotic formula, which should be superior to the Hasofer-Lind approximation for the ease of small failure probability instances, the latter depending not only on the shortest distance to the boundary surface, but also on the curvatures of this surface at the design point. Here we evaluate simple examples from the probabilistic strength of materials, following Elishakoff and Hasofer (1989).

A.2 Beam Subjected to a Concentrated Force

Consider the beam shown in Fig. A.3, loaded by concentrated force P with the following (deflection) failure criterion

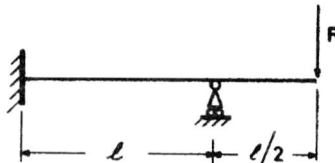

Fig. A. 3 A beam under concentrated force

$$u_{max} = \frac{5}{48}\frac{Pl^3}{EI} \geq \frac{1}{30}l \qquad (A.5)$$

where the span l *is* a deterministic constant (in this case 5 m), while the modulus of elasticity E and moment of inertia I as well as the applied force P, are uncorrelated random variables with the following characteristics

$$\bar{P} = 4\ kN, \qquad \sigma_p = 1\ kN$$

$$\bar{E} = 2\times10^7\ kN/m^2, \qquad \sigma_E = 0.5\times10^7\ kN/m^2 \qquad (A.6)$$

$$\bar{I} = 10^{-4}\ m^4, \qquad \sigma_I = 0.2\times10^{-7}\ m^4$$

The limit state function (or safety margin) reads

$$g(P,E,I) = \frac{1}{30}l - \frac{5}{48}\frac{Pl^3}{EI} \qquad (A.7)$$

or, substituting the numerical value of l,

$$g(P,E,I) = EI - 78.12P \qquad (A.8)$$

We are interested in the probability of failure

$$P_f = Prob(EI - 78.12\ P < 0) \qquad (A.9)$$

A.3   Approximate Solutions

Expanding the "safety margin" M defined as

$$M = EI - 78.12P \qquad (A.10)$$

in a Taylor series in the vicinity of the mean values $E = \bar{E}, I = \bar{I}, P = \bar{P}$ we find

$$\bar{M} = \bar{E}\bar{I} - 78.12\bar{P} = 1687.52\ kNm^2 \qquad (A.11)$$

$$\sigma_M^2 \cong \left[\left(\frac{\partial M}{\partial E}\right)_0 \sigma_E\right]^2 + \left[\left(\frac{\partial M}{\partial I}\right)_0 \sigma_I\right]^2 + \left[\left(\frac{\partial M}{\partial P}\right)_0 \sigma_P\right]^2 \qquad (A.12)$$

where the zero subscript signifies that the partial derivatives were evaluated at the mean value $\bar{E}, \bar{I}, \bar{P}$, so that

$$\sigma_M = 645.06\ kNm^2 \qquad (A.13)$$

The approximation of the reliability index β according to the mean-value first-order second-moment method is thus

$$\beta = \frac{\bar{M}}{\sigma_M} = 2.616, \quad P_f = \Phi(-2.616) = 0.00044 \qquad (A.14)$$

Let now E, I, and P constitute independent normal variables. Then $\bar{M}$, given by Eq. (A.11), is an exact value of the mean safety margin, whereas for σ_M an exact value can be found, namely

$$E(M^2) = E\left(E^2 I^2 - 2\times 78.12 EIP + 78.12^2 P^2\right)$$
$$= \left(\sigma_E^2 + \overline{E}^2\right)\left(\sigma_I^2 + \overline{I}^2\right) - 2\times 78.12\overline{EIP} + 78.12^2\left(\sigma_P^2 + \overline{P}^2\right) \qquad (A.\ 15)$$
$$\sigma_M^2 = E\left(M^2\right) - \overline{M}^2 = \sigma_E^2 \overline{I}^2 + \sigma_I^2 \overline{E}^2 + 78.12^2 \sigma_P^2 + \sigma_E^2 \sigma_I^2$$

so that $\sigma_M = 640\ kNm^2$, whence

$$\beta = \frac{\overline{M}}{\sigma_M} = 2.64\ , \qquad P_f = \Phi(-2.64) = 0.000554 \qquad (A.\ 16)$$

For their part Thoft-Christensen and Baker (1982), using the Hasofer-Lind method, found $\beta = 3.29$ and $P_f = 0.00071$. Let us compare this value with an exact solution given below.

A.4 Exact Solution

We first determine the probability density function of the margin of safety, given $E = e_0$. The random variables $\{M|E = e_0\}$ is also normal, with mean

$$E\left(M|E = e_0\right) = e_0 \overline{I} - 78.12 P = \mu(e_0) \qquad (A.\ 17)$$

and mean-square deviation

$$\sigma(e_0) = \sqrt{e_0^2 \sigma_I^2 + 78.12^2 \sigma_P^2} \qquad (A.\ 18)$$

The conditional probability density is then

$$f_{M|e_0}\left(m|E = e_0\right) = \frac{1}{\sigma(e_0)}\phi\left[\frac{m - \mu(e_0)}{\sigma(e_0)}\right] \qquad (A.\ 19)$$

and its unconditional counterpart

$$f_M(m) = \int_{-\infty}^{\infty} \frac{1}{\sigma(e_0)}\phi\left[\frac{m - \mu(e_0)}{\sigma(e_0)}\right]\frac{1}{\sigma_E}\phi\left(\frac{e_0 - \overline{E}}{\sigma_E}\right)de_0 \qquad (A.\ 20)$$

Finally, the desired probability of failure is

$$P_f = Prob(M < 0) = \int_{-\infty}^{0} dm \int_{-\infty}^{\infty} \frac{1}{\sigma(e_0)}\phi\left[\frac{m - \mu(e_0)}{\sigma(e_0)}\right]\frac{1}{\sigma_E}\phi\left(\frac{e_0 - \overline{E}}{\sigma_E}\right)de_0 \qquad (A.\ 21)$$

However

$$\int_{-\infty}^{0} \frac{1}{\sigma(e_0)}\phi\left[\frac{m - \mu(e_0)}{\sigma(e_0)}\right]dm = \int_{-\infty}^{-\mu(e_0)/\sigma(e_0)} \phi(z)dz = \Phi\left[-\frac{\mu(e_0)}{\sigma(e_0)}\right] \qquad (A.\ 22)$$

Therefore

$$P_f = \frac{1}{\sigma_E}\int_{-\infty}^{\infty}\Phi\left[-\frac{\mu(e_0)}{\sigma(e_0)}\right]\phi\left(\frac{e_0 - \overline{E}}{\sigma_E}\right)de_0 \qquad (A.\ 23)$$

or, with the new variable

$$\frac{e_0 - \overline{E}}{\sigma_E} = t, \tag{A. 24}$$

we obtain

$$P_f = \int_{-\infty}^{\infty} \Phi\left[-\frac{\mu(e_0)}{\sigma(e_0)}\right]\phi(t)dt, \quad e_0 = t\sigma_E + \overline{E} \tag{A. 25}$$

With our numerical values, we find $P_f = 0.00064$, and defining $\overline{\beta} = -\Phi^{-1}(P_f)$, we find it to equal 3.205. At first glance, the value of 0.00044 for P_f (Eq. (A.14)) may seem a not too bad approximation compared with its exact counterpart, but this cannot generally be said of the mean-value first-order method as is demonstrated in the following design problem.

A.5 Design of Structural Element

Consider, following Toft-Cristensen and Baker (1982) the beam shown in Fig. A.4, loaded with a concentrated P. Deterministic analysis yields for the maximum deflection, in standard notation,

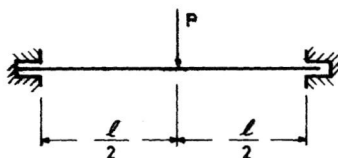

Fig. A. 4 A clamped-clamped beam under concentrated random force

$$u_{max} = \frac{1}{192}\frac{Pl^3}{EI} \tag{A. 26}$$

Again, the span l is a deterministic constant (equals 6 m), whereas the others are uncorrelated random variables with the following characteristics:

$$\overline{P} = 4\ kN, \quad \sigma_p = 1\ kN \tag{A. 27}$$

$$\overline{E} = 2\times10^7\ kN/m^2, \quad \sigma_E = 0.5\times10^7\ kN/m^2$$

The mean value $\overline{I}$ of the moment of inertia is unknown and the problem is to determine $\overline{I}$ so that $P_f = 0.00135$ or $\beta = 3$, failure being associated with the inequality

$$u_{max} \geq \frac{1}{100}l \tag{A. 28}$$

and moreover

$$\sigma_I = 0.1\overline{I} \tag{A. 29}$$

The failure surface reads

$$g = \frac{1}{100}l - \frac{1}{192}\frac{Pl^3}{EI} = 0 \qquad \text{(A. 280)}$$

or

$$6EI - 112.5P = 0 \qquad \text{(A. 31)}$$

Note that in the book by Thoft-Christiansen and Baker (1982) this equation is given with a typographical error as $6EI - 1113P = 0$

The basic variable P, E and I are normalized as $Z_1 = (P - \bar{P})/\sigma_P$, $Z_2 = (E - \bar{E})/\sigma_E$ and $Z_3 = (I - \bar{I})/\sigma_I$. In the normalized coordinate system the failure surface is given by

$$6 \times 10^7 (2 + 0.5Z_2)\bar{I}(1 + 0.1Z_3) - 112.5(4 + Z_1) = 0 \qquad \text{(A. 32)}$$

The design point is now given by $z_i = \beta\alpha_i = 3\alpha_i$. The unknowns μ_I, α_1, α_2 and α_3 and are determined by the set of equations

$$\mu_I = \frac{112.5(4 + 3\alpha_1)}{6 \times 10^7 (2 + 1.5\alpha_2)(1 + 0.3\alpha_3)}$$

$$\alpha_1 = \frac{1}{k}112.5 \qquad \text{(A. 33)}$$

$$\alpha_2 = -\frac{1}{k}10^7 (3 + 0.9\alpha_3)\bar{I}$$

$$\alpha_3 = -\frac{1}{k}10^7 (1.2 + 0.9\alpha_2)\bar{I}$$

and k is determined by the condition

$$\alpha_1^2 + \alpha_2^2 + \alpha_3^2 = 1 \qquad \text{(A. 34)}$$

As explained earlier, the above set can be solved iteratively, choosing starting values for β, α_1, α_2, and α_3, and calculating the new values by Eq. (A.33) until the process converges (Table A.1). Note that, the starting values are the same as in the paper by Thoft-Christiansen and Baker (1982), but because of the discrepancy in the equation for the failure surface, the final value is $\bar{I} = 165.474 \times 10^{-7} m^4$ instead of $\bar{I} = 167 \times 10^{-7} m^4$ as given by Thoft-Christiansen and Baker (1982).

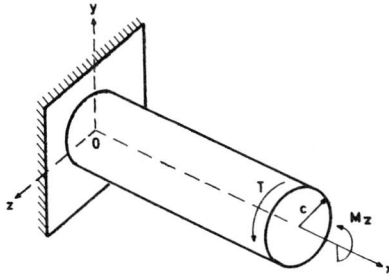

Fig. A. 5 A shaft under random bending moment and random torque

Table A.1: Iteration needed to find the structural reliability

		ITERATIONS					
	Start	1	2	3	4	5	6
$\bar{I} \times 10^7$	40	115.3	153.3273	114.4085	165.374	165.4615	165.747
α_1	0.58	0.35503	0.245577	0.2283007	0.2269919	0.2267891	0.226778
α_2	-0.58	-0.901712	0.9606184	-0.9671366	-0.967689	-0.967738	-0.967742
α_3	-0.58	-0.144336	-0.112514	-0.109884	-0.109773	-0.1097546	-0.109753

As regards the mean value first-order method, we first determine the mean and mean-square values of the safety margin, which turn out to be

$$\overline{M} = 6\overline{EI} - 112.5\overline{P} = 12 \times 10^7 \, \overline{I} - 0.450 \qquad (A.\ 35)$$

$$\sigma_M \cong \sqrt{\left(6\overline{I} \times 0.5 \times 10^7\right)^2 + \left(6 \times 2 \times 10^7 \times 0.1\overline{I}\right)^2 + 112.5^2} \qquad (A.\ 36)$$

Solving the quadratic equation

$$\frac{12 \times 10^7 \, \overline{I} - 450}{\sqrt{10.44 \times 10^{14} \, \overline{I}^2 + 112.5^2}} = 3, \qquad (A.\ 37)$$

we find

$$\overline{I} = 207.30924 \times 10^{-7} \, m^4 \qquad (A.\ 38)$$

which is more than 20% above the value to the Hasofer-Lind method. The conclusion is that caution should be exercised in using the mean value first-order even where small fluctuations of the design variables are involved. This mean value first-order is adopted in a number of textbooks without warning the reader of the possible errors involved (see, e.g. Kapur and Lamberson (1977)). On the other hand the Hasofer-Lind method yields good accuracy, as will be shown by contrasting it with the exact solution as follows.

We are to find $\bar{I}$ and $\sigma_I = 0.1\bar{I}$ such that the probability of failure has a specific value.

$$\text{Prob}(6EI - 112.5P < 0) = \text{Prob}(M = EI - 18.75P < 0) = 0.00135 \qquad (A.\ 29)$$

Consider again, as in the preceding section, a conditional random variable $\{M|E = e_0\}$, which is normal with mean and mean square deviation

$$E(M|E = e_0) = e_0\mu - 75 \qquad (A.\ 39)$$

$$\sigma(e_0) = \sqrt{0.01e_0^2\mu^2 + 351.5625} \qquad (A.\ 40)$$

The appropriate conditional probability density is then

$$f_{M|e_0}(m|E = e_0) = \frac{1}{\sigma(e_0)}\phi\left[\frac{m - E(M|E = e_0)}{\sigma(e_0)}\right] \qquad (A.\ 41)$$

The unconditional probability becomes

$$f_M(m) = \int_{-\infty}^{\infty}\frac{1}{\sigma(e_0)}\phi\left[\frac{m - E(M|E = e_0)}{\sigma(e_0)}\right]\frac{1}{\sigma_E}\phi\left(\frac{e_0 - \bar{E}}{\sigma_E}\right)de_0 \qquad (A.\ 42)$$

The probability of failure is then obtained, after some algebra, in analogy to the preceding section

$$P_f = \int_{-\infty}^{\infty}\frac{1}{\sigma_E}\phi\left(\frac{e_0 - \bar{E}}{\sigma_E}\right)\phi\left[-\frac{E(M|E = e_0)}{\sigma(e_0)}\right]de_0 \qquad (A.\ 43)$$

Finally, Eq. (A.39) takes the form

$$P_f = \int_{-\infty}^{\infty}\phi(t)\Phi\left[\frac{75 - 0.5(t + 4)x}{\sqrt{0.0025(t + 4)^2 x^2 + 351.5625}}\right]dt = 0.00135 \qquad (A.\ 44)$$

where

$$x = \bar{I}\times10^7 \qquad (A.\ 45)$$

The process of tackling the transcendental Eq. (A.45) is shown in Fig. A.5., where the solution is plotted against x. The value of x at which the probability of failure equals 0.00135 is 167.38, which exceeds by about one percent the value yielded by the Hasofer-Lind method.

Appendix B
Biographical Notes

In this chapter the biographical notes are provided about some engineers and scientists who played the central role in either originating the idea of the safety factors (Coulomb, Navier), or combining the concept of safety factors and the notion of probability (Kazinczy, Mayer, Mukhadze, Freudenthal, Kakushadze, Rzhanitsyn, Streletskii). The probability inequalities by Bienaymè-Markhov and Chebychev were used by several engineers to derive bounds on probability of failure in the safety factor context. The biographical notes of these prominent mathematicians are also included, clearly showing that mathematics and engineering are inseparable. Notes are put in the alphabetical order of the authors.

I-J. Bienaymé

Irenée-Jules Bienaymé was born on August 28, 1796 in Paris, France. In 1815 he enrolled in the École Polytechnique, which was dissolved because of political reasons. He was employed as a translator for journals; in 1818 he became lecturer of mathematics at the military academy at St. Cyr, leaving in 1820. In 1820 he enters the Administration of Finances as an inspector, and in 1834 he was elevated to the rank of the Inspector General. Because of revolution of 1848, he decided to retire and dedicate his remaining years to scientific work. He became a member of French Academy of Sciences in 1852. In 1853 he derived what is now known as Bienaymé (or Bienaymè-Markhov) inequality $Prob\ [|X\text{-}E(X)| \geq \varepsilon] \leq Var(X)/(\varepsilon^2 n)$. Chebychev obtained the inequality in 1867 (in a much more restricted setting and with a more difficult proof), in a paper published simultaneously in Russian and French, juxtaposed to reprinting of Bienaymé's paper. Later, Chebychev gave Bienaymé credit for arriving at the inequality via the "method of moments", whose discovery he ascribes to Bienaymé.

I-J. Bienaymé passed away on October 19, 1878 in Paris, France.

P.L. Chebychev

Pafnutii Lvovich Chebychev was born on May 26, 1821 in Okatovo (Kaluga region), Russia. As E. Seneta writes in the book by Johnson and Kotz (1997), "In 1847, Chebychev began to teach at St. Petersburg University, eventually becoming full professor in 1860, in which year he took the course in probability theory (on the retirement of V. Ya. Buniakovsky), which reawakened his interest in the subject area. He had been promoted to the highest academic rank of the St. Petersburg Academy of Sciences in the previous year."

"He was a leading exponent of the Russian tradition of treating the probability calculus as an integral part of mathematical training, and through the "Petersburg mathematical school," of which he was the central figure, his overall influence on mathematics within the Russian empire was enormous."

"… the year 1858 seemed to mark the beginning of a mutual correspondence [between Bienaymé and Chebychev] and admiration between the two men, leading to the eventual election of each to a membership in the other's Academy of Science. Even though in 1874 Chebychev gave Bienaymé credit in print for arriving at the inequality via "the method of moments," whose discovery he ascribed to Bienaymé, and this view was later reiterated by Markov, it is a valid point that it was more clearly stated and proved by Chebychev. In any case, through the subsequent writings of the strong Russian probability school, Chebychev's paper has undeniably had the greater publicity, to the extent that the inequality has often born Chebychev's name alone."

P. L. Chebychev passed away on December 8, 1894, in St. Petersburg, Russia.

Ch. A. de Coulomb

Charles Augustin de Coulomb was born on June 14, 1736 in Angoulême, France. Timoshenko (1953) writes: "After obtaining his preliminary education in Paris, he entered the military corps of engineers. He was sent to the island of Martinique where, for nine years, he was in charge of the various works in construction which led him to study the mechanical properties of materials and various problems of structural engineering." Straub (1952) informs: "…he used the opportunity [of being as a Génie officer to the French colony of Martinique] to investigate the solidity and statical behaviour of building elements, especially walls and vaults, and to deal with the problem mathematically. The results of these investigations which were, at the outset, merely destined for his personal use, were summarized as a treatise which he submitted to the Académie des Sciences and which appeared in the 1773 volume of the "Mémoires des Savants Etrangers", under the title "Essais sur une application des régles de maximis et minimis à quelques problèmes de statique relaifs à l'architecture."

Gillmor (1971) writes in the book dedicated to Coulomb:

> "In this one memoir of 1773 there is almost an embarrassment of riches, for Coulomb proceeded to discuss the theory of comprehensive rupture of masony piers, the design of vaulted arches, and the theory of earth pressure. In the latter he developed a generalized sliding wedge theory of soil mechanics that remains in use today in basic engineering practice. A reason, perhaps, for the relative neglect of this portion of Coulomb's work was that he thought to demonstrate the use of variational calculus in formulating methods of approach to fundamental problems in structural mechanics rather than to give numerical solutions to specific problems."

According to Straub (1952):

> "as far as the development of the methods of structural analysis is concerned, it is mainly the above-mentioned treatise from his Martinique period which must be regarded as a document of vital significance. In it, the classical problem of the bending of a beam has been fully solved, exhaustively and correctly, and even the problem of shearing stress is touched upon. In the same treatise Coulomb also develops a thesis for the failure of compression stressed masonry and brick structures...
>
> The publication of Coulomb's *Mèmoire* ought to be regarded as a milestone in the history of structural analysis. Unfortunately, the rich contents of the treatise were compiled in such a concise form and concentrated into so little space that, as Saint-Venant points out, most of it escaped the notice of experts for 40 years. This was all the more understandable as the author, in later years, occupied himself no longer with these problems, but turned to other branches of physics...
>
> ...Rather incidentally, in connection with his observations on strength tests (tension, shearing and bending tests), Coulomb also deals, in his *Mèmoire,* with the bending problem of a cantilever beam of rectangular cross-section. His treatment is universal in that he takes into consideration the shearing strength as well as the compressive and tensile strength, and admits, in principle, of any relationship between stress and strain. As a special case, applicable to the perfectly elastic body, he then obtains the well-known relation $M=\sigma bh^2/6$."

After his return to France, Coulomb worked as an engineer at Rochelle, the Isle of Aix, and Cherbourg. In 1779, he shared (with Van Swinden) the prize awarded by the Academy for a paper on the best way to construct a compass; 1781 saw him win the Academy prize for his memoir "Théorie des machines simples", in which the results of his experiments on the friction of different bodies slipping on one another...were presented. After 1781, Coulomb was stationed permanently in Paris, where he was elected to membership of the Academy... He turned his attention to researches in electricity and magnetism" (Timoshenko, 1952). Kragelski and Schedrov (1956) wrote: "Coulomb's contributions to the science of friction were exceptionally great. Without exaggeration, one can say that he created this science." "Between 1802 and 1806... he was inspector general of public instruction, and, in that role, he was mainly responsibly for setting up the lycées across France" (O'Connor and Robertson, 2003).

Heyman (1972) points out about Coulomb:

> "He was not interested in 'applied mathematics' but in the use of mathematics to obtain solutions to actual practical problems."

Coulomb passed away on August 2, 1806 in Paris, France.

A. M. Freudenthal

Alfred Martin Freudenthal was born in Poland on February 12, 1906. As Professor Harold Liebowitz (1976) – then the Editor-in-Chief of the journal *Engineering Fracture Mechanics* wrote: "Professor Freudenthal belongs to that small group of scientists and engineers who originally came from the region of the Carpathian Mountains of the former Austro-Hungarian Monarchy. Following its disintegration after World War I, and the resulting destabilization of Central Europe that culminated in the Nazi nightmare, many people fled the area. Being thus "fission products," members of this group have displayed a characteristically strong interaction with their new surroundings, as demonstrated by the quality, number and effect of their technical contribution, and the wide range of their scientific interests. Born in 1906, Freudenthal spent his childhood in various parts of Austria where his father, Simon Freudenthal, a civil engineer served with the Austrian Technical Railroad Administration. The end of World War I in 1918 found the family in Austrian Silesia, a region fought over between the newly formed states of Poland and Czechoslovakia and partitioned between them in 1920. Finishing his secondary school education on the Polish side, he entered the Technical University of Lwow [now in Ukraine] in 1923, transferring a year later to the German Technical University to study with Prof. J. Melan, the most distinguished bridge designer of his time in Central Europe. Freudenthal was awarded a degree in civil engineering in 1929 in Prague, and in 1932 in Lwow. In 1930 he was awarded the degree of Doctor of Technical Sciences by the German Technical University in Prague on the basis of his dissertation of plasticity."

"He started his professional career in 1930 as a structural designer with a well-known consulting engineering firm in Prague, returned to Poland in 1934 to collaborate for a short time with Prof. M. T. Haber in Warsaw and emigrated in 1935 to Palestine (Israel), then a territory under British Mandate. He became the chief structural engineer and subsequently the resident engineer in the planning, construction, and technical administration of a new port in Tel Aviv between 1936 and 1946. In 1937 he accepted an appointment as lecturer, later as a professor of bridge engineering at the Hebrew Institute of Technology in Haifa."

"Participating in 1937 from Palestine in an international competition for the design of a new bridge over the Vistula in Warsaw, he won third prize. During World War II he served as consultant to the Chief of Engineering, British Forces, Palestine and Transjordan, and was in charge of the construction of the minesweepers for the British Navy which were launched and commissioned in the port of Tel Aviv for service in the Pacific area."

Professor Tichý informs that he has "an unconfirmed information that A. M. Freudenthal was for a short period in 1938 or so employed with the Klokner Institute of the Czech Technical University of Prague."

"In 1947, on the basis of a paper on the statistical aspects of fatigue, sponsored by H. J. Gough for publication in the *Proceedings of the Royal Society (London)*, he was invited to visit the United States and to lecture at several Universities. Upon recommendation of the late Prof. H. F. Moore, he accepted an appointment with the University of Illinois as Visiting Professor of Theoretical and Applied Mechanics."

Between the years 1949 to 1969 he held an appointment as Professor of Civil Engineering at Columbia University. In 1969 he joined the School of Engineering and Applied Science of the George Washington University as Professor of Civil and Materials Engineering and Director of the Institute for the Study of Fatigue and Structural Reliability.

A. M. Freudenthal has been awarded the Norman Medal of the American Society of Civil Engineering in 1948 and 1957, and the von Karman Medal in 1972. In 1956 he was awarded the Medal of the Royal Swedish Aeronautical Society. He was elected a member of the American Academy of Engineering. The American Society of Civil Engineers has established the Freudenthal Medal "in honor of his outstanding accomplishments in research, teaching, and engineering practice"; it is awarded biannually to an individual in recognition of distinguished achievement in the area of safety and reliability applied to civil engineering.

His first paper "Safety and Working Stresses" (in Hebrew) on probabilistic mechanics was published in 1938, in the *Journal of Associates of Engineers of Palestine*. His first paper in English on this topic became a classic; it is entitled "Safety of Structures" (1947). It appears to be a must-read for every probabilistic mechanicist.

A. M. Freudenthal passed away on September 27, 1977.

Harold Liebowitz wrote: "Professor Freudenthal was one of the seminal engineers and scholars of his era." In the foreword to the book *Selected Papers* by Alfred M. Freudenthal published in the series "Civil Engineering Classics" by the American Society of Civil Engineers, the members of The Engineering Mechanics Division's Ad-hoc Committee on the A.M. Freudenthal Volume, Messr. Gerard Fox, Robert Heller, Harold Liebowitz, Paul Parisi, M. Shinozuka and W.R. Spillers wrote: "He was a man of extreme breadth – a genius if you like – ranging at will over both theory and practice in engineering. Even as we sit with his work in front of us the enigma deepens."

A.M. Kakushadze

Editorial of the journal *Stroitelnaya Mekhanika i Raschet Sooruzhenii* (*Structural Mechanics and Analysis of Building Constructions*) in its issue No.3, 1974 writes: "Alexander Moiseevich Kakushadze was born on November 17, 1903. In 1929 he graduated from the Faculty of Physics and Mathematics in Tbilissi State University [in Georgia], and the civil engineering faculty of Transcaucasus Institute of Transportation."

"… Starting 1929 he worked in the departments of theoretical mechanics and higher mathematics in the Transcaucasus Industrial Institute. Later he became a Dean of the Georgian Polytechnic Institute." Since 1963 he served as a chairman of the

270

Department of Strength of Materials and the Theory of Elasticity at the Georgian Polytechnic Institute (presently Technical University).

"Over 20 years he directed the section of structures, theory of constructions and experimental station of Georgian Scientific Institute of Buildings and Hydroenergy ..."

"He authored over 100 scientific works, including eight monographs. Academician N. Muskhelishvili gave the following assessment of the A. Kakushadze's scientific activity: "Numerous works of Alexander Kakushadze are of both theoretical and important practical values. One should especially emphasize the ability of Professor Kakushadze to use complex mathematical apparatus for the solution of concrete engineering problems. Professor Kakushadze is a scientist of great erudition and scope who significantly contributed to the theory and practice of solution of numerous engineering problems." Inter alia, he co-authored with Professor G. Mukhadze the monograph about applications of mathematical statistics for establishment of coded values and the safety factor of building materials and structures." According to Losaberidze (1984), "Professor Kakushadze was one of the first scientists [in the Former Soviet Union] who paid an attention to the possibility of the utilization of the methods of mathematical statistics and theory of probability in structural mechanics."

The books by Streletskii (1947) and by G. Mukhadze and A. Kakushadze (1954) are the only ones up to now whose titles mention the concept of safety factor.

A. Kakushadze passed away on July 28, 1981 in Tbilisi, Georgia.

G. Kazinczy

In his review paper, Streletskii (1963) writes: "Understanding of variability of the circumstances of the works of structures is *a priori* clear ... The desire to take into account this variability appeared already in 1913 in the works of a Hungarian [scientist] Prof. Kazinczy, who suggested statistical method of accounting the variability."

Kaliszky (1984) provided the biographical data: "Gábor Kazinczy was born in Szeged, Hungary, on January 19, 1889. He studied at the Technical University, Budapest and graduated as a civil engineer in 1911. As a young engineer he entered municipal service, mainly engaged in checking structural design of new constructions in Budapest. In 1931 he obtained the Doctoral degree from the Technical University, Budapest. In 1939 he was appointed associate professor..."

"In 1913 — in course of evaluating a load test — he made the significant observation that the load carrying capacity of a built-in rolled steel beam was not exhausted by reaching the yield moment at one cross section (*Betonszemle*, No. 4, 5, 6, 1914). He was of the view that as a result of the yield moment in the cross sections so-called plastic hinges developed. The load carrying capacity of the complete beam is not exhausted until it is not entirely transformed by successively developing hinges to a chain of hinged bars. This realization and its explanation made Kazinczy the founder of a discipline the limit analysis of structures."

Of importance to us is the note made by Kaliszky (1984): "During World War I, he was engaged with safety problems but he did not publish his results earlier than 1929, at a Congress in Vienna."

Kaliszky (2003) informs, relating to the above remark made by Streleskii (1963): "Professor Lenkei studied the life and scientific activity of Kazinczy and informed me... that in 1928 at the Conference of the "Internationale Vereining für Brücken und Hochbau (IVBH)" (International Association for Bridge and Structural Engineering, IABS, in English) he had a discussion on Gehler's lecture on safety in the presence of Streletskii. It is possible that this discussion referred to a paper published by Kazinczy around 1913."

Kaliszky (1984) writes: "Prof. Kazinczy's wide range of interests and abilities is seen from his concerns, in addition to theoretical and experimental analysis of structures, also in material testing, fire safety, labor safety, as well as in site management, technology and economy problems of constructions."

Complete list of his publications appeared in the editorial published in the journal *Acta Techn. Acad. Sci. Hung*, Vol. 55, pp. 455-460, 1966.

Gábor Kazinczy passed away on May 26, 1964 in Sweden.

M. Mayer

Max Mayer was born in Salzburg, Austria on September 16, 1886. During the years 1926 to 1930 he was a Professor at the State Civil Engineering University in Weimar. He authored several books, all in the German language:

(a) "Wirtschaftlichkeit als Konstrukzionsprinzip im Eizenbetonbau" (Economy as Principle of Construction of Buildings with Reinforced Concrete), 1913.
(b) "Die Anregung Taylors für die Baubetrieb" (Suggestions of Taylor for Building Activities), 1915,
(c) "Betriebswissenschaft Handbibel für Bauengineurs" (Science of Industrial Organization), 1926.
(d) "Die Sicherheit der Bauwerke" (Safety of Buildings), 1926.
(e) "Nomographie der Bauingenieurs" (Nomography in Civil Engineering), 1927.
(f) "Zweifel Bei der Gestaltung von wasserdichten Kellern" (Uncertainties Concerning the Construction of Waterproof Cellars), 1937
(g) "Neure Statik der Tragwerke aus biegesteiffen Stäben" (New Statics of Structures with Nonflexible Beams), 1937.
(h) "Die Statische Berechnung" (Static Analysis), 1953.

His 1926 book titled *Sicherheit der Bauwerke und Ihre Berechnung nach Grenzkräften anstatt nach zulässigen Spannungen* is a central one on the context of safety factors. In 1975 Technical Institute of Materials and Constructions (INTREMAC), Madrid, Spain, published its translations into English and Spanish. Professor Milik Tichý wrote an introduction. It appears most instructive to reproduce it nearly fully:

"A long time experience of the mankind shows that important ideas never arise solitarily; many confirmations of this fact can be traced particularly in the history of technology and natural sciences."

"Another fact, which also is well known and proved, is that outstanding ideas with a potentially strong impact on the future fall into complete oblivion and are covered by the darkness of ignorance or conservatism of those people who would be expected to carry on the development of thoughts."

"It is not difficult to detect the causes of these two phenomena. The advancements of technology and human knowledge, and, simultaneously, the development of economy, form a fertile ground for step-by-step accumulation of something what we may call "quanta of thoughts." Then, as soon as a certain amount of quanta is collected, the new idea is suddenly formulated. Since the changes of technology and economy are worldwide, it is very likely that the critical point will be reached at several places almost simultaneously."

"However, the human thinking is in general more advanced than the possibilities which are offered by the contemporaneous technological and economical situation. In other words: new seed needs fresh soil. If the soil is not prepared, the seed merely vegetates and often is covered by dust of oblivion."

"All this is also valid for one of the most important innovations in the field of structural design: in the probabilistic approach to the structure and loads, and in the economical assessment of structures in sociological entities."

"During the past few years three extremely interesting publications have been discovered which were completely unknown to scholars working in the field of probabilistic limit states design, which however contain a surprising amount of modern attitudes to the design of structures."

"Carl Forssell in 1924 formulated, with a certain caution, but firmly, the notion of the economical determination of safety levels and the notion of acceptance of a certain reasonable risk in construction. Of course, at that time Forssell's theoretical equipment could not be sufficient for a thorough treatment of his reasoning."

"Fifty years elapsed now since Max Mayer from Duisburg, Germany, published a… [book] in which he examined practically all problems discussed now in various structural safety committees… What is perhaps the most important result of Mayer's work, is a working proposal of a probabilistic limit states code for various types of building materials, attached to the main paper."

"Finally, in 1929, N. Khotsialov from Kichkas, [former] Soviet Union, proposed a method of determining design quantities based on probabilistic and economical considerations. He did not assume the randomness of load, as Forssell and Mayer did, but on the other hand he was conscious of the important influence of the quality control upon the design parameters. He proposed a system of assessment of failures, inspection, and consequent standardization which has not been reached in any country yet. His paper merits to be republished ..."

"Evidently, the three authors never met; they lived in countries with different economical situations, but they set foundation stones for a large intellectual structure. Unfortunately, these stones were hidden by deposits of indifference. Maybe the new thoughts were felt to be too audacious even by their authors themselves since none of them, as far as it is known, developed his ideas further. Their brilliant successors, M. Prot, N. S. Streletskii, W. Wierzbicki, R. Lévi, A. Freudenthal, A. G. Pugsley and many others had to build up the structure again, this time, however, being supported by the accelerating development of science, technology and economy."

Bolotin (1965) writes: "The works of M. Mayer, N. F. Khotsialov and N.S. Streletskii were substantially the first works in the domain of reliability theory. A number of questions of reliability theory were first formulated and solved in these works, a circumstance which merits being noticed in modern publications on reliability theory."

M. Mayer passed away on July 29, 1967 in Starnberg.

The biographical information herein was obtained from the books "Kürschners Deutscher Gelehrten-Kalender (1928/29, 1950, 1961, 1971), and private communications by Zimmermann (2003) and Tichý (2003).

G.M. Mukhadze

Giorgi M. Mukhadze was born 29 November 1886 in Tbilisi, Georgia. His father passed away when he was five years old. The age of eight he was accepted to the three year church school, upon graduation from which he entered Tbilisi gymnasium for counts, despite his peasant origin. In 1906 he entered the civil engineering department of the Tomsk Technological Institute in Russia, which he interrupted several times to gain practical experiences. Since 1916, he is back to Georgia as an engineer of the building department and the city architect in Tbilisi. Since 1918 he teaches at the Railroad Engineering College, while since 1928 he teaches at the Georgian Industrial Institute, where he became a Professor in 1931. Later he became a chairman of the Department of Structural Mechanics (1933-1963), and served as the Dean of Civil Engineering Faculty during the years 1933-1942. In 1946 he was elected as a Corresponding Member of the Georgian Academy of Sciences and for 15 years led the Department of Theory of Constructions. His first scientific work was published in 1931; he authored about 40 papers, including two books.

According to Academician Sekhniashvili et al (1986), "the research subject of Giorgi Mukhadze constituted classical problems of structural mechanics and the methods of their solution…in that period the modern computer techniques were unavailable and it is easy to visualize how much effort and time were needed for solutions of complex problems… G. Mukhadze's goals were development of universal and at the same time simple methods," such as graphical statics, replacement of the method of fictit load method and others. Academician Oniashvili notes, "G. Mukhadze almost always and elegantly utilizes geometric methods." One of the methods developed by Professor G. Mukhadze he characterized as a "jewel of structural mechanics."

In 1952 Professor Giorgi Mukhadze and Professor Aleksander Kakushadze had a foresight to write a book on the statistical justification of the safety factors. The only other book that was exclusively devoted to this topic was that by Streletskii (1947). As Rzhanitsyn (1978) noted Streletskii was unable to obtain a correct relationship between safety factor and reliability. The first rigorous analyses were due to Freudenthal (1947) and Rzhanitsyn (1947). The book by Mukhadze and Kakushadze (1954) used Rzhanitsyn's (1947) derivation of the safety factor-reliability relation.

Professor G. Mukhadze passed away on January 5, 1963 in Tbilisi, Georgia.

L.M.H. Navier

Claude Louis Marie Henri Navier was born on February 10, 1785 in Dijon, France. Straub (1952) writes in his book on a history of civil engineering: "Louis Marie Henri Navier (1785-1836), whose name is familiar to engineers through the theory, named after him, of the consistently plane cross-sections, was born four years before the outbreak of the French Revolution. From 1802 to 1807, he studied at the Paris "Ecole Polytehnique" which had been founded in 1794 under the auspices of Monge and Carnot, … and subsequently at the "Ecole des ponts et chaussées." Then, only 22 years of age, he started his practical career as "Ingénieur des Ponts et Chaussées" of the Seine Département."

"Apart from practical engineering works (among his works are several bridges over the Seine, e.g. those at Choisy, Asnières, Argenteuil), Navier revealed an early inclination for scientific and pedagogic activities. Soon after the conclusion of his own studies, he joined the teaching staff of the "Ecole des ponts et chaussés," first as a deputy assistant, then in 1821 as an extraordinary professor for Applied Mechanics. In this capacity, he regarded it as his main task to apply the discoveries and methods of theoretical mechanics to practical tasks of construction, and to equip the engineering students with a appropriate scientific armour."

"…Navier for the first time integrated the isolated discoveries of his predecessors in the field of applied mechanics and related subjects into a single, unified system of instruction, and that he taught his students how to apply the laws and methods already known to the practical tasks of structural engineering, i.e. to the determination of structural dimensions. In doing so, he became the actual creator of that branch of mechanics which we call *building statics*, or *structural analysis*."

"... As far as the theory of flexure is concerned, the problem had already been solved by Coulomb fifty years earlier ... but Coulomb's success had not become known to a wider circle of engineers. Even Navier's first publications do not give the correct solution; they still subscribe to Bernoulli's and Mariotte's assertion that the position of neutral axis is indifferent."

"But in 1819, Navier corrects his error partly so that he arrives at correct results for symmetric cross-sections. But it is only in 1824 that he formulates the modern terms (correct where stress and strain are proportional) for the deflection and ultimate strength of the beam subject to bending."

F. Stüssi (1940) sums up the importance and particularity of Navier's work in the following words: "The task which Navier set himself, is nothing less than the formulation of a proper method of structural analysis ... The fact that we are able, today, to construct safely and economically, is mainly due to the methods of structural analysis, that particular branch of mechanics which is based on the actual working conditions of a structure. These methods were created, within little more than a decade, by a single man, Navier."

C. L. M. H. Navier passed away on August 21, 1836 in Paris, France.

A note on Navier, written by another French giant of mechanics, Saint-Veriant, is reproduced in the book on a history of theory of structures by Charlton (1982).

A.R. Rzhanitsyn

Aleksei Rufovich Rzhanitsyn was born on August 28, 1911.

The journal *Stroitelnaya Mekhanika i Raschet Sooruzhenii* (*Structural Mechanics and Analysis of Building Constructions*) wrote (issue 4, 1976; issue No. 1, 1982; issue No.1, 1988):

"A.R. Rzhanitsyn was a student of Moscow Building Engineering Institute during the years 1931 to 1936 ... In 1945 he was awarded the Ph.D. degree. Since 1936 he worked in the Central Scientific Institute of Buildings (CNIPS), which was re-organized in 1957 into CNIISK. In this institute he worked until 1982. Since 1953 he served as the Chair of the strength of materials department. Since 1972 A.R. Rzhanitsyn served as a Professor in the Department of Structural Mechanics at the Moscow Civil Engineering Institute (MISI)."

"For the best scientific works A.R. Rzhanitsyn was awarded the Galerkin prize three times."

"There is no part of the modern structural mechanics in which A.R. Rzhanitsyn did not bring considerable contribution."

"Much attention was devoted be A.R. Rzhanitsyn to development of probabilistic methods of analysis of civil constructions. Already in 1947-1949 he took into account the statistical nature of loads and the properties of the material for determination of the safety factor; he obtained approximate formulas, that take into account probability of collapse during repeating loading."

"In the statistical method suggested by him in 1949-1971 the eccentricity of the load application and the initial curvature were treated as random variables..." "In the monograph *Theory of Analysis of Structures and Reliability*, published in 1978, the systematic exposition of the reliability theory was given and methods of reliability based design were developed." A.R. Rzhanitsyn authored ten monographs and over 170 scientific works.

A. R. Rzhanitsyn passed away in 1987 in Moscow, Russia.

Bolotin (1965) writes: "The study of the statistical nature of the safety factor was continued on a broader front in the post-war [WW II] years. The investigations by A. R. Rzanitsyn occupied an important role."

N.S. Streletskii

Nikolai Stanislavovich Streletskii was born on September 15, 1885. The following represents quotes from the editorial published in the journal *Stroitelnaya Mekhanika i Raschet Sooruzhenii* (*Structural Mechanics and Analysis of Building Constructions*) (issue No.1, 1985). "In 1911 he graduated from the St. Petersburg Institute for Transport Engineering. Tens of bridges of different systems, including the Old and New Dnepr, "Moskovoretskie" bridges and others were built with his participation and direction. During years 1933-1937 he serves as the director of the Central Scientific Institute of Civil Constructions (presently CNIISK)."

"Since he ascribed fundamental importance to the realistic description of the structural behavior, N.S. Streletskii constantly devoted his time to the problems of strength and reliability, to the justified choice of the safety factor. His investigations in this field became scientific basis for the development of the analysis method via the limit states... As a result in 1955 fundamental revision of design codes took place. His pedagogical activity started in 1915 in Moscow High Technical College (presently Technical University), where he headed the Department of Metal Structures. He authored over 250 scientific works."

Professor Tichý informs that "the first result of Streletskii were put in life in 1949 by a decision of the Supreme Soviet of the USSR."

On January 31, 1931 he was elected as a Corresponding Member of the Division of Mathematical and Natural Sciences, of the Academy of Sciences.

Belenia (1975) writes: "The work "Basics of Statistical Accounting of Safety Factors in Structures" is one of the fundamental ones, in which author [N.S. Streletskii] developed the method of theoretical determination of the safety factor via the principle of equal strength, based on the theory of probability and methods of statistics."

Iliasevich (1975) writes: "He devoted over 15 scientific works to the problem of safety factors, allowable stresses and collapse of structures, justifiably considering it extremely important for development of metal structures. In these he gave a deep, analytical and clear review of all mentioned problems, that are most important for the entire engineering enterprise. During the World War II when the lack of the metal was acutely felt, N. S. Streletskii, using his investigations on safety factors and allowable stresses justified the security and feasibility of considerable increasing of stresses in metal structures, which was practically implemented."

Bolotin (1965) writes: "N. S. Streletskii had an outstanding role in the development of statistical methods in structural mechanics."

N. S. Streletskii passed away on February 15, 1967.

Author Index

Bolotin V.V., 1, 26, 29, 51, 95, 97, 133, 145, 199, 203, 208, 229, 246, 272, 276
Bompas-Smith J.H., 208, 209
Borch K.H., 209
Borges J.F., 209
Bornstein A.E., 214
Bosovich R.G., 43, 230
Bougund U., 211
Bouton I., 209
Breitung K., 209, 257
Breneman J.E., 204
Breugel K., van 209
Brown J.A.C., 64
Brown K.A.P., 209
Brown C.B., 209
Broding W.C., 209
Brown C.B., 41, 209
Bruevich N.G., 209
Bruhn E.F., 48, 209
Bucher C.G., 209
Bulychev A.P., 209, 210
Buniakovsky V.Ya., 264
Burros R.H., 210
Bushnell D., 143, 144, 210
Bussiere R., 210
Butterfield B., 210
Bychawski Z., 237

Cable C.W., 210
Camp B.H., 180, 181, 210
Campbell C.C., 210
Capitanio R.S., 228
Carpenter R.B., Jr., 210
Carpenter R., 213
Carter A.D.S., 1, 118, 198, 199, 210
Casciati F., 6, 26, 39, 40, 47, 210, 249
Castanetha M., 209, 218
Castellani A., 210
Catelanni G., 10
Cathey B.H., 210
Cavalier T.M., 173, 174, 234
Cawley J.C., 211
Cederbaum G., 211
Ceisser S., 217

Cempel C., 211, 236
Cesare M.A., 211, 252
Chamis C.C., 10
Chay S.C., 234
Chazen C., 246
Chebychev P.L., 153, 157, 159, 163, 164, 165, 211, 216, 217, 264, 265
Checkanov N., 248
Cheen W.K., 211
Chen G.H., 211
Chen K., 252
Chen Y.M., 211
Cheng K.S., 231
Chertykovtsev V.K., 204
Chilver A.H., 211
Chiras A.A., 229, 246
Chirkov V.P., 10
Chou K., 211
Church J.D., 45, 211
Churchley A.R., 211
Clarck G.M., 173, 174, 238
Close E.R., 211
Cohen H., 211
Collins K.R., 12, 237
Colombi P., 140, 216
Cooke R., 141, 206, 212
Coombs C.F., 227
Coquhon I., 212
Cormier D., 228
Cornell C.A., 1, 16, 41, 58, 142, 199, 200, 202, 207, 212, 217, 220, 230, 255
Corotis R.B., 211
Corso J.M., 27
Coulomb C.A., 6, 9, 42, 43, 44, 220, 264, 265, 266, 273
Courrian P., 228
Cramér H., 185, 187, 188, 212
Craswell K.J., 237
Cross N., 212
Crowley V.F., 212
Cruse V.F., 212
Curback M, 212
Curnick G.E., 212

Subject Index

294

Other Books Authored or Co-Authored by Prof. I. Elishakoff

"Probabilistic Theory of Structures" (Wiley and Sons, 1983; Second edition by Dover Publications, 1999).

"Convex Models of Uncertainty in Applied Mechanics" (Elsevier, 1990; with Y. Ben-Haim).

"Random Vibrations and Reliability of Composite Structures" (Technomic, 1992; with G. Cederbaum, J. Aboudi and L. Librescu).

"Probabilistic and Convex Modeling of Acoustically Excited Structures" (Elsevier, 1994; with Y. K. Lin and L. P. Zhu).

"Non-Classical Problems in the Theory of Elastic Stability" (Cambridge University Press, 2001; with Y. W. Li and J. H. Starnes, Jr.).

"Finite Element Methods for Structures with Large Stochastic Variations" (Oxford University Press, 2003; with Y. J. Ren).

"Eigenvalue Problems of Inhomogeneous Structures: Unusual Closed-Form Solutions" (CRC Press, 2004; to appear).

Presently he is preparing the monograph *"Follower Forces: Fact or Fancy?"*.